The Open University

A Second Level Course

SYSTEMS BEHAVIOUR

Module 6

The structure and management of ecosystems

Prepared for the
Systems Behaviour Course Team
by Dick Morris

THE OPEN UNIVERSITY PRESS

The Systems Behaviour Course Team

R. J. Beishon (*Chairman*)
M. Amos (Course Assistant)
G. S. Einon (Biology)
D. A. Elliott (Technology)
E. S. L. Goldwyn (BBC)
J. Groom (BBC)
G. E. Harland (Staff Tutor)
L. M. Jones (Systems)
R. Jones (BBC)
R. D. R. Kyd (Editor)
J. O. N. Lawrence (Staff Tutor)
R. McCormick (Institute of Educational Technology)
R. M. Morris (Systems)
J. J. Naughton (Systems)
D. Nelson (BBC)
G. Peters (Systems)
A. R. Thomas (Systems)
B. Whatley (BBC)
P. I. Zorkoczy (Electronics)

Consultants

W. Fincham (Queen Mary College)
P. Mottershead (National Institute of Economic and Social Research)

Revision Team

R. J. Beishon (Systems)
D. Bolton (Systems)
S. Brown (Systems)
R. Carter (Systems)
D. A. Elliott (Technology)
L. M. Jones (Systems)
R. D. R. Kyd (Editor)
R. M. Morris (Systems)
J. J. Naughton (Systems)
G. Peters (Systems)
C. Pym (Course Assistant)

Consultants

D. Campbell
N. Lees (Victoria Infirmary, Glasgow)
J. Marshall (University of Bath)
R. J. Mills (Centre for Respiratory Investigation)
S. Montgomery (University College, London)
P. Mottershead (National Institute of Economic and Social Research)

The Open University Press, Walton Hall, Milton Keynes

First published 1973. Second edition 1977

Designed by the Media Development Group of the Open University

Produced in Great Britain by
Technical Filmsetters Europe Limited, 76 Great Bridgewater Street, Manchester M1 5JY.

This text forms part of an Open University course. The complete list of modules in the course appears at the end of this text.

For general availability of supporting material referred to in this text, please write to the Director of Marketing, The Open University, PO Box 81, Milton Keynes MK7 6AT.

Further information on Open University courses may be obtained from the Admissions Office, The Open University, PO Box 48, Milton Keynes MK7 6AB.

2.1

The Open University Systems Behaviour Course

This course is designed to introduce students from many different backgrounds – social scientists, technologists, scientists – to the inter-disciplinary study of systems and their behaviour. It is a second level course leading to a half-credit qualification.

The course is presented in eight 'modules' each of which contains study material for about two weeks. This includes a printed text, two television programmes, a radio programme and, at certain points, experiments and computer-based activities. Each module consists of a detailed case study of a particular system, followed by a presentation of some analytical tools or techniques.

The case studies are:

deep-sea container port operation,
air traffic control at Heathrow Airport,
industrial social systems,
local government,
the British telephone system,
grassland ecosystems,
the human respiratory system,
a shipbuilding firm.

The techniques provide methods for:

collecting and handling data,
modelling and simulating systems behaviour,
analysing data statistically.

Students completing the course will have sufficient knowledge of eight different systems to analyse and understand their behaviour. Further, they will have acquired a set of concepts, approaches and ideas which are fundamental to understanding the behaviour of any complex system.

All flesh is grass . . .

Isaiah 40, v 6.

CONTENTS

Foreword

Interest in ecosystems has increased considerably during the last decade. Concern for the environment over pollution and the depletion of natural resources has produced a range of reactions from the hysterical to the resigned. In any course on systems it is desirable to include material on ecosystems, but the problem we faced was how to choose an example for a case study which was factual and well-researched and also one which revealed the system's behaviour fairly clearly. We decided to concentrate on what might be described as a rather prosaic case, that of sheep production, because this demonstrates well the complexity of an apparently straightforward farming system. We were fortunate in having on the course team Dick Morris, who, as an applied ecologist, had done fundamental research in this area himself.

As with all our case studies in this course, we are asking you to grapple with yet more specialist terms and concepts, but we have tried to balance the range of these against the learning load involved. As you will appreciate by now, each subject area generates its own terms and ideas, often rather unkindly called 'jargon'. I hope you can see that in most cases these new words are an efficient 'shorthand' which, once you become reasonably familiar with it, can help a good deal in understanding what is going on. We hope that, even if you regard yourself as 'non-mathematical', you are coming to see that mathematical formulations are in some senses a similar form of shorthand, with the added advantage that the symbols and concepts can be combined and manipulated in a very powerful way to arrive at useful equations or expressions which assist in understanding the behaviour of systems.

This module shows very clearly just how useful mathematical formulations can be in an area you may have regarded as basically non-mathematical. It also develops further the idea of modelling a system by mathematical expressions and relations. It introduces the important techniques of difference and differential equations, and matrix methods.

I should like to emphasize the importance of this case study and module because it demonstrates very clearly two things: first, the value of the systems approach in directing attention to the many important factors which might be overlooked or regarded as irrelevant because they belong to a different discipline or field of study; second, the value of the scientific approach and its attempts to quantify the variables and to identify the relationships between them in mathematical form. If you are unfamiliar with this approach, you will certainly find it difficult, but it is well worth persevering with the material because, unless you are able to follow the steps of beginning to model and quantify systems variables, it will be difficult to go beyond producing relatively simple descriptions of a system's behaviour.

This is the first biologically based system with which we have dealt. Remember to think as you work through the material about the differences between biological systems and the constructed man-made systems dealt with earlier in the course.

John Beishon

Aims

1 To show how systems methods can be applied to the study of systems comprising whole living organisms and their environments.

2 To compare these systems with others studied earlier in the course, and to develop further techniques which can be used generally for the study of systems.

3 To show how these techniques can be used to increase understanding of the functioning of living systems, to predict their behaviour and to plan their management.

Objectives

After reading this module you should be able:

1 To define what is meant by an ecosystem.

2 To describe the components and processes involved in the function of an ecosystem; autotrophs and heterotrophs; photosynthesis, respiration and synthesis; growth, reproduction, death, nutrient intake and waste output.

3 To appreciate the complexity of ecosystems.

4 To define population dynamics and energy flow.

5 To construct a compartment model of a simple ecosystem.

6 Given suitable information, to quantify such a compartment model in terms of difference or differential equations and to understand the principles of solving such equations.

7 To set up a transition matrix as a model of a population of organisms.

8 To describe in principle the validation of models of ecosystems.

9 To describe how sheep meat and wool are produced in the United Kingdom.

10 To describe the interactions between the sheep production system in the United Kingdom and other systems.

Preface

It would help you to have a copy of the book *The Biosphere** and Spedding's book *Grassland Ecology*† while reading this module, though these are not essential. You will find chapters 1, 3, 5, 7 and 9 of *The Biosphere* most useful after you have read section 1 of this module. You will be referred to specific parts of Spedding's book in sections 4–7.

Before reading any of the material in this module, you should have read quickly Dale's article 15 'Systems analysis and ecology' (pp. 254–68) in the *Systems Behaviour* Reader.‡ Do not worry if some of the terms are unfamiliar. Return to it and re-read the article carefully after working through this module; you will find it gives a survey of the material in this module.

I am grateful to Professor C. R. W. Spedding, formerly of the Grassland Research Institute, Hurley, for many of the ideas in this module, and to his colleagues Jean Walsingham and Bob Large for giving me access to many of their unpublished data.

*Scientific American (*1970*) The Biosphere, *Freeman.*

†*Spedding, C. R. W. (1971*) Grassland Ecology, *Oxford University Press.*

‡*Beishon, R. J., and Peters, G. (eds.) (1976) Systems Behaviour (2nd edn), Harper & Row/The Open University Press.*

1 Introduction

1.1 Definition of the term 'ecosystem'

Previous modules of this course have looked at systems comprising men and/or machines. In this module, we are going to look at systems comprising whole organisms – that is, whole *plants*, whole *animals* (such as man), *bacteria*, etc. – which, together with parts of the surroundings in which they live, make up that class of systems called *ecosystems*. These have been defined by a New Zealand ecologist, R. B. Miller, as

> open systems, comprising plants, animals, organic residues, atmospheric gases, water and minerals, which are involved together in the flow of energy and the circulation of matter.

ecosystem

Evidently they are *natural* systems; but they may also include elements of *human-activity* systems.

Their most important property is that they are open systems, that is, *open* to exchanges with their environment. We shall look first at what this means for ecosystems in general, and then in more detail at their structure and function. Following that, we shall develop some of the techniques which we can use to study and, hopefully, to understand and manage these systems.

open system

1.2 Ecosystems and entropy

As implied in Miller's definition, ecosystems include two major classes of component: *living organisms* and certain *non-living materials*. This distinction may appear obvious, but it involves some problematic cases, such as viruses, which possess some characteristics of both living and non-living matter, and plant leaves or other organs which may be dead but are still part of the parent organism.

living organism
non-living material

The most fundamental distinction between living and non-living components is that living things are highly organized throughout their structures, whereas non-living things, although they may be closely ordered on a microscopic scale, are generally *less organized*. (Crystals are highly organized, but crystalline substances in bulk are generally composed of a disorganized mass of small crystals.) The degree of organization of any object can be measured in terms of its *entropy*. Any object which is highly organized, with a well-defined structure, is said to have low entropy. Disorganization represents high entropy. In any real system, entropy always *tends* to increase.

entropy

> To ask why this is so takes us beyond the scope of this module, into the theory of thermodynamics. An introduction to the theory of entropy is given in the Open University *Science Foundation Course*, Unit 5 The States of Matter if you wish to delve more deeply into this aspect. Bertalanffy's article no. 2 in the *Systems Behaviour* Reader also uses the concept of entropy, though you will probably not find this very helpful.

high entropy   low entropy

In a closed system, this increase in entropy continues until it reaches a stable value, that of maximum possible *disorder*. In an open system, this trend can be counteracted by the input of 'order' in some form from outside the system. Entropy and energy are inter-related, and the tendency for entropy to

entropy and energy

increase in a system can be countered by taking in energy from outside. This is what happens in the open systems of this module, ecosystems.

The non-living part of an ecosystem comprises mainly simple, unorganized chemical compounds, such as soil minerals, water, and gases in the air.

SAQ 1

SAQ 1

Would these have high or low entropy?

Alone, these non-living components would form a closed system, so there would be nothing to prevent their entropy increasing to its maximum value. In contrast, living organisms are complex and highly organized, and can maintain their high organization by taking in energy. We can regard them as machines which take in energy and use that energy to create and maintain *order*, contrary to the universal tendency of an increase in *disorder*. Through the activity of the living organisms, the overall level of order in the ecosystem is also maintained or even increased.

1.3 Ecosystems in outline

How is this reduction in entropy accomplished? The living components of ecosystems are divided into two major classes; these are:

1 *Autotrophs* – green plants, algae (such as the *Chlorella* in your Home Experiment Kit), some bacteria, etc.

autotroph

2 *Heterotrophs* – animals (the *Daphnia* used in your home experiment), fungi, etc.

heterotroph

(The suffix *-troph* means feeding. *Auto*trophs are self-feeding and *hetero*trophs feed *on* other organisms.)

Energy enters the whole ecosystem through the autotrophic components, to be used to counter entropy increase. Autotrophs take in energy largely by the process of *photosynthesis*. Green plants and some other organisms possess special pigments which can trap light energy from the sun, and they can use this energy to convert carbon dioxide and water from the non-living part of the ecosystem into sugar, which they retain in their bodies. Carbon dioxide and water have relatively simple molecules, with low* energy and hence a high entropy value. Sugar is a much more complex, organized compound, of higher energy and lower entropy.

photosynthesis

Here is a simplified formulation of the conversion from carbon dioxide to sugar for those who are familiar with chemical 'shorthand',

Carbon
dioxide + water + energy → sugar + oxygen
$$6CO_2 + 6H_2O + X \text{ joules} \rightarrow C_6H_{12}O_6 + 6O_2$$

| low-energy molecules | | high-energy molecule | low-energy molecule |

(A joule is a unit of energy.)

A by-product of this reaction is another low-energy molecule, oxygen. This is of vital importance to heterotrophs, including ourselves, who need oxygen to breathe. Heterotrophs cannot use solar energy directly, but they can use energy which has been trapped in the autotrophs, by feeding *on* them. The energy which was held in sugars in autotrophs is then released by a process

* Strictly, this refers to the free energy of the molecules. A more detailed treatment of energy in chemical reactions is given in The Open University (1974) The Man-made World: Technology Foundation Course Units 24/25 Chemical Technology, The Open University Press.

which is effectively the reverse of photosynthesis, called *tissue respiration.* (Sometimes this is also called internal respiration, to distinguish it from external respiration, which is the process of breathing.)

SAQ 2

Suggest a similar 'shorthand' representation of tissue respiration.

The energy released from the sugar is used by the heterotrophs either for movement, warmth, etc. or for the organization of simpler compounds in their food into the complex low-entropy components of their bodies, such as fats, proteins and bone. This process of tissue respiration is common to all living organisms: autotrophs such as plants are clearly not composed entirely of sugars, so they, too, use some of the energy which was initially trapped in sugar for the formation of proteins and other low-entropy compounds from simple substances taken in from their surroundings. None of these conversions is perfectly efficient. Thus the difference between the energy stored in low-entropy (high-energy) components of the living organisms and the energy in the high-entropy (low-energy) compounds from which they were made is less than the energy needed actually to make them. Suppose there are two simple substances A and B, which have 20 joules and 10 joules of energy stored in them. From these, to form a third compound C, which can store 50 joules of energy, we need a supply of more than $50-(20+10)=20$ joules. This excess of energy is needed to overcome biochemical 'friction' in the conversion system, and is lost, as heat energy, back to the surroundings and, ultimately, out into space.

The high-energy, low-entropy compounds in living organisms continually tend to break down into their simpler components, with consequent release of energy and the return of these simple components back to the surroundings. For organisms to retain their integrity, they must replace these complex compounds as they break down, so they are continually importing energy in some form. This may either be a direct gain from sunlight by autotrophs, or as high-energy living material eaten by heterotrophs.

SAQ 3

List the components through which energy may pass, from its initial source, through an ecosystem.

In an ecosystem we have seen that the underlying structure involves the uptake of energy and its conversion by *photosynthesis* into high-energy sugars in the bodies of *autotrophs*. This energy is released by *tissue respiration*, either within autotrophs or within *heterotrophs* feeding on autotrophs. The energy released is used for the *synthesis* of other high-energy, low-entropy compounds from simple substances of higher entropy.

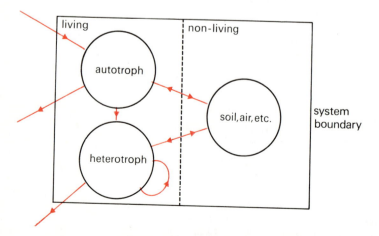

Figure 1   Flow diagram for energy in a simplified ecosystem

We can summarize this situation very neatly by a technique with which you should be familiar, the flow diagram. Figure 1 shows the flow of energy through the major components of an ecosystem. One point of interest is the flow which leaves the heterotroph component and returns to it: this emphasizes that not only do heterotrophs take energy from autotrophs, but that they also get it from other heterotrophs.

Exercise

Construct a similar flow diagram for the flow of materials such as sugars.

See Figure 2. Note that in this diagram all the material flows remain within the ecosystem. This may not be true for all ecosystems; it depends where the system boundary is defined. Remember that the boundary of a system depends very largely on the observer's involvement with that system, as emphasized in earlier modules. We shall return to the problem of defining boundaries in ecosystems in the case study.

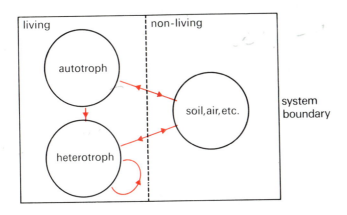

Figure 2   Flow diagram for material in a simplified ecosystem

We now see in broad terms how the flow of energy (or entropy) and material occurs in ecosystems. The actual circulation of materials involves five major processes. These are covered in the next section.

1.4   Material flow in ecosystems

The five major processes by which material is moved around an ecosystem are:

1   Nutrient intake,

2   Waste output,

3   Growth,

4   Reproduction,

5   Death.

1.4.1   *Nutrient intake*

This is defined as including all those processes where material is taken into the body of a living organism, as in the eating of one animal by another, the uptake of carbon dioxide by a leaf or of water by a root. Here is a list of terms describing various methods of taking in nutrients.

Diffusion,

Predation,

Herbivory,

Saprophytism,

Decay,

Parasitism,

Symbiosis,

Respiration.

See how many you can define. The definitions are given in Table 1 overleaf.

Table 1   Definitions of forms of intake

Diffusion. Process whereby fluids (liquids or gases) pass through a porous membrane, or across a space, as in the diffusion of gases into the leaves of plants, or of solutions into their roots.

Predation. Killing and eating of one animal by another (Figure 3).

Figure 3   Predation: predatory lions kill and eat other animals (usually herbivores) such as this zebra

Herbivory. Eating of a living plant by an animal (Figures 4 and 5).

Saprophytism. Consumption of a dead plant or animal by fungi or certain plants.

Decay. Breakdown and consumption of dead plant or animal by bacteria or fungi.

Parasitism. Process whereby one organism lives in or on another and gains its food supply via the host organism with detrimental effects on the host.

Figure 4   Herbivory: cows grazing

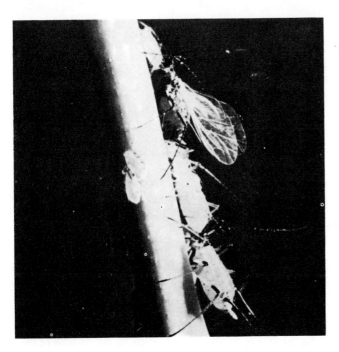

Figure 5   Herbivory: aphides on a plant

Symbiosis. Process whereby two organisms live in close proximity, each 'feeding' the other to mutual benefit.

Respiration (a trick!). This refers to breathing: the intake of oxygen by a plant or animal – a physical process as opposed to the biochemical process of tissue respiration.

1.4.2   *Waste output*

Material passes into the body of an organism by an intake process. Not all this material is useful; for example, the tougher parts of a piece of grass eaten by a sheep simply pass straight through the animal and emerge as faeces. Some material is broken down into simpler compounds of no further use, and these are eliminated from the body. For example, proteins are ultimately broken down into urea, which is eliminated in urine. Similarly, the carbon dioxide formed during tissue respiration is lost through the lungs of an animal, or through the small pores (stomata) in the leaves of plants. (This process is also referred to as respiration!)

We have grouped together all such processes as waste output. A list of processes is given in Table 2.

Table 2   Definitions of different waste-output processes

Respiration.	Removal of carbon dioxide from the body of an animal or plant.
Transpiration.	Removal of water from plant.
Perspiration.	Loss of water from animal.
Guttation.	Another process of water loss from plants.
Urination.	Removal of soluble nitrogenous wastes from animals.
Defaecation.	Removal of wastes from the gut of animals.
Moulting, sloughing, shedding.	Loss of parts of body of growing organisms (exhibited by birds, insects. *Daphnia*, deciduous trees, etc.)

15

1.4.3 *Growth*

The balance between nutrient intake and waste output is the amount retained in the body of an organism. This is used for two visible processes, growth and reproduction, and also for maintenance purposes, to repair normal wear and tear. Growth generally involves adding to the total amount of material within the original organism, as in the conventional sense of an overall increase in size. It can also involve an increase in size of a specified organ within the body, or a change in composition, as when a twig increases in strength as it becomes more woody.

1.4.4 *Reproduction*

For most organisms there is a finite limit to their size.

Exercise

Why do you think this is?

Once an organism reaches its size limit, material which would otherwise be used for growth can then only be used either to maintain its existing structure, or to produce entirely new individuals. This is the process of *reproduction*. It may take place either sexually or asexually (as in the vegetative propagation of plants by cuttings), but in each case results in the formation of one or more new, smaller organisms which can then grow to the (limited) size typical of its particular type.

It depends primarily on the size of the parent organisms, that is to say, it is genetically controlled. This genetic control ensures that the organism does not exceed limits imposed by the strength of its supporting tissue – for example, bones – by its need for mobility or by the sort of place in which it lives.

1.4.5 *Death*

A living organism lives by taking in material and energy from outside itself, that is, it is an open system and can counter the normal process of increasing entropy. When it loses this ability, it begins to become *disorganized*, and finally reaches a stage which is termed death. There are problems in defining exactly what is meant by death, but at its most fundamental level, it implies the inability to decrease or even maintain the current level of entropy in that organism. Past this point, the living organism becomes part of the non-living environment, though it may still have a high enough level of energy to act as a food source for saprophytes or decay organisms (see Table 1).

death

SAQ 4

Summarize the components of an ecosystem, the processes responsible for energy transfer and the processes responsible for material transfer.

SAQ 4

A pictorial summary of ecosystem components and processes appears in Figure 6.

At this point you may like to read chapters 1, 3, 5, 7 and 9 of *The Biosphere*. You do not need to spend long on this; all you require are a few more details of the functioning of ecosystems in general, to put some flesh on the skeletal outline given here.

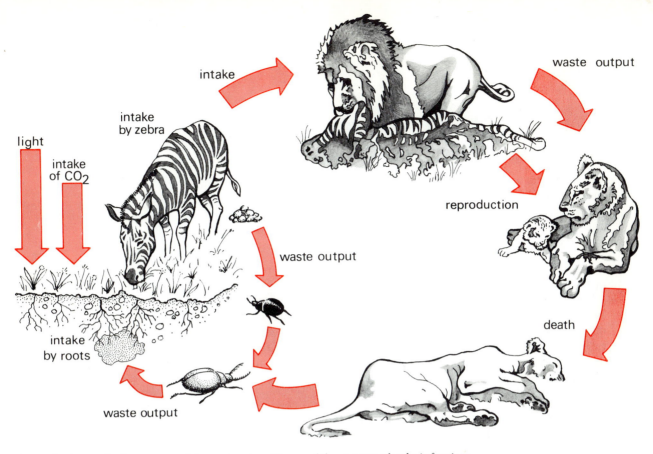

Figure 6   A generalized ecosystem and its components, with some of the processes whereby it functions

2 Approaches to the study of ecosystems

2.1 The problem under study

The preceding section has given an outline of the basic functioning of all ecosystems. The detailed picture, as you will see from the case study, is much more complex, since any one ecosystem will contain not just a single autotroph and a single heterotroph, but a large number of different species of each, and large numbers of individuals within each species.

A typical lawn, which is a relatively simple system, will contain probably four or five different species of grass, with up to about 2000 individuals of each species in every square metre, not to mention the inevitable weeds. In addition, there will be between twenty and a hundred different species of soil insects, snails, millipedes, etc., with up to 1000 individuals of each species in each square metre of soil, and several million bacteria. Each of these will be involved in its own energy and material flow processes, exchanging several hundred different chemical compounds with other organisms and with the environment.

The potential complexity of behaviour in such a system is tremendous. If you were a student of the Technology Foundation Course, you may remember the number of possible states exhibited by a small number of light bulbs, as mentioned in Unit 1 Systems. Out of this, ecologists are attempting to produce some sort of order, proceeding according to the normal phases of scientific inquiry. These are:

1 Observation and classification of events,

2 Experimentation,

3 Synthesis into a coherent, predictive theory.

The study of ecology is currently largely in the second phase of this series. Unfortunately, pressures from pollution and a rapidly expanding world population on a finite natural world need remedies which can only be developed from coherent predictive theory. This lack of such a body of theory may be part of the reason for some of the wilder statements which are made in the name of 'ecology' and 'the environment' at the moment.

SAQ 5

How do these three phases of scientific investigation correspond with the phases given by Dale for a systems study on pp. 255–62 of the *Systems Behaviour* Reader?

SAQ 5

2.2 The development of ecological research

The careful *observation and classification* of the *static properties* of ecosystems began with the earliest civilizations, who identified different species of animals, plants, etc. This was put on a systematic basis in the eighteenth century by the Swedish botanist Linnaeus, whose terminology is still used. The identification of different species can be a simple process or it can be very difficult, depending on the apparent similarity or otherwise of different species of organism. Thus an elephant and a cow may be obviously different species, but what about two different species of grass? On p. 47 you will be asked to look at some grass plants in detail. When you do this, see if you can pick out obviously different species within the site you choose. These static properties (i.e. properties which do not change with time) are relatively simple to classify, though the process may be tedious.

observation and classification
static properties

When it comes to the *dynamic properties* of ecosystems, classification is much less easy. The numbers, rates of growth, size, etc. of the organisms present will change from day to day within an ecosystem, and may even change from hour to hour. Remember how many different organisms there may be in a simple system like a lawn. Even to observe all these changing quantities becomes a major problem. Attempts to classify, experiment on, and predict changes become almost impossibly difficult. Here we have a classic 'systems' problem. How has it in fact been tackled? So far, there have been two general approaches to such problems, called the population-dynamic and energy-flow approaches.

dynamic properties

In a study of *population dynamics*, interest centres on the numbers of individuals of each species present, and the interactions between the numbers within different species. For example, many experiments have been conducted to determine how and why the numbers of various insect species change under specified conditions, or with changes in these conditions.

population dynamics

In a study of *energy flow*, experiments trace the path of energy through the ecosystem. The idea is to determine how much energy enters the system through plants and to follow this through the heterotroph and other components of the system to discover where it is either stored or finally leaves the system. Although the names of the species involved will almost certainly be recorded, their number, and factors controlling these numbers are of secondary interest to their total energy. This approach clearly has a close relationship to the function of an ecosystem as an energy-exchange system, but does not give a complete picture any more than does a study of population dynamics alone. Remember that ecosystems not only involve the transfer of energy, but also of materials, which include individual organisms.

energy flow

material flow

Systems techniques are designed to deal with complex collections of interacting entities, a category which clearly includes ecosystems. Can we use such techniques to provide a comprehensive picture and an understanding of ecosystem function? The next section attempts to answer this question.

2.3   Systems contribution to the study of ecosystems

2.3.1   *Systems questions*

You should by now be familiar with the series of questions which you would ask if you were investigating any system. A summary of the relevant questions for an ecosystem follows:

1   Why are we studying the system?

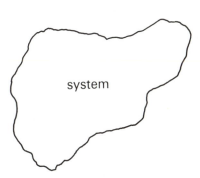

2   Where does the boundary of the system lie, and what is its environment? (Dale's lexical phase.)

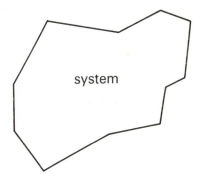

3   What are the components and subsystems within the system, and how are they inter-related?

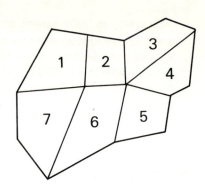

4   What are the input and output variables associated with the subsystems and the system as a whole?

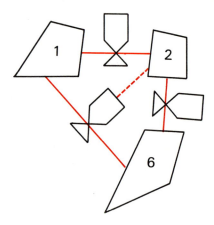

5   What fixed properties do the components and subsystems possess? (These are often referred to as the *parameters* of the subsystems; this terminology will be discussed later.)

parameter

(3–5 are Dale's parsing phase.)

6   How do we measure the variables associated with the system, and how do these variables change in response to normal and abnormal inputs? (Dale's modelling phase.)

7   How is the system likely to develop in the future?

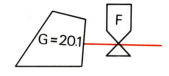

SAQ 6

How does this list differ from the list given in Module 1 Deep-sea Container Ports?

SAQ 6

Stages 2–6 of this list involve making a *model* (but note Dale's terminology) of the system under examination and possibly using it to *simulate* the behaviour of the system. In terms of the history of scientific progress given earlier (section 2.1) this modelling exercise is the first stage in the synthesis of the classification and experimentation into a more general body of *theory*. A model may be used to predict the behaviour of a particular system in response to a particular input. From a study of numerous models, it may be possible to arrive at a more general theory, which explains, or at least predicts, the behaviour of a whole class of systems. This is the phase which Dale calls analysis.

model and theory

How do we apply the series of questions posed above to ecosystems? In the following sections we shall try to answer each question in turn, and then we shall go on to develop some specific techniques which are useful in modelling ecosystems.

2.3.2   *Why do we study ecosystems?*

Man depends for his existence on the interactions between whole organisms. They supply his food, maintain the oxygen content of the atmosphere and break down many of his noxious waste products. Clearly, it is important to understand how the whole 'life-support system' works, and to understand how it is likely to respond either to novel inputs or to changes in its subsystems. In particular, man may be concerned to preserve a species of animal which he finds attractive (usually, of course, man himself) or he may wish to maintain some complete landscape. He may wish to maximize the production of some particular foodstuff or of some natural product like wool or rubber.

Thus man can set objectives for an ecosystem which he is attempting to control. Note that this is not an objective *of* the ecosystem itself, but is something which man attempts to impose on such a system, of which he is not always a part. The ecosystem itself does not have any corporate goal, objective or ideal.

In fact, the ecosystem as a *system* is a product of the man who studies it; the components of the ecosystem are unaware that they belong to any system. This contrasts ecosystems with all the other systems studied earlier in this course, which all either comprised men, or were designed by men. A weakness of Ackoff's terminology is revealed when we try to apply his concepts to ecosystems. This is why questions 2 and 3 from Module 1 were omitted from our list in section 2.3.1. Although the ecosystem itself has no goal or objective, this may not be true of the individual organisms within that system. Each organism in an ecosystem is probably trying to obtain the maximum possible growth rate for itself, and to reproduce at the maximum possible rate within the limits set by all the other organisms in the system. We shall return to this concept when we discuss the control of ecosystem processes in section 3.2.1.

2.3.3 *Boundaries of ecosystems*

Defining the boundaries of an ecosystem, as with other systems, reflects the interests and aims of the person setting these boundaries. The only complete ecosystem is the whole biosphere, since there are no absolute physical and biological boundaries which can be placed round smaller entities. However, we may well want to regard a defined area or a defined agricultural enterprise as an ecosystem, within Miller's definition given in section 1.1. In the studies of animal populations which have been conducted over many years in Wytham Wood near Oxford, the investigators have defined the boundaries of the system as being the physical boundaries of the wood itself. In the case-study part of this module, we look at grassland sheep production, defining the boundaries in terms of this business enterprise.

The important lesson which should have emerged from our course is to take care always to include all relevant components within the system, and to identify all the inputs and outputs which may cross the defined boundaries. This is no less important for an ecosystem than for any other system.

2.3.4 *Components and subsystems*

The next step is to identify the components of the ecosystem. We have already done this at the most general level.

SAQ 7

List the major ecosystem components.

SAQ 7

At a more detailed level, each of the major subsystems might be divided into different classes, such as carnivores (flesh eaters), large herbivores (plant eaters), small herbivores, soil animals, herbs, shrubs and trees, depending on their size or position within the ecosystem. Figure 7 illustrates some examples of ecosystem components.

Alternatively, we might identify *functional subsystems*, such as the soil–water system, which we might regard as *black boxes*. These are defined only in terms of the relationships between their inputs and outputs, although they may be identifiable with particular physical parts of the system. The components of each subsystem might be identified as particular species, or even parts of organisms, such as leaves on a tree, although most ecologists prefer to direct their attention to whole organisms.

functional subsystem

black box

21

Figure 7(a) Carnivore: larva of great diving beetle (Dytiscus marginalis) *feeding on tadpole*

Figure 7(b)   Herbs: bluebells

Figure 7(c)   Shrub: hawthorn

Figure 7(d)   Tree: elm

2.3.5 *Input and output variables*

The inputs and outputs from system components all change as the system *behaves*, so they are referred to as *variables*. This is a term you will already have come across in the Mathematics Booklet or will be familiar with from your mathematical knowledge. In fact, our use of the term is borrowed from mathematical usage. Typical variables might be the *numbers* of a specified organism present, the rate of growth of a crop or the *level* of some

variable

contaminant present in the ecosystem. Output variables for the system as a whole might be measures of its *stability* (its lack of change with time), its *diversity* (effectively, the number of different species present) or its *productivity* (a measure of the total amount of energy trapped in the whole system per unit of time).

The number of variables which could be used to describe an ecosystem's behaviour is large. We have already mentioned two common classes of measurements on ecosystem variables.

SAQ 8

SAQ 8

What were these two common measurements?

In order to answer our systems questions up to this point, we shall have to have built up a picture of the inter-relationships between the various subsystems; this may have been in the form of a flow diagram or simply a verbal description. Both of these are *models* of the system being studied. So far, however, they would be unquantified. To make the model more useful, we have to specify the relationships between the systems variables.

2.3.6 *Systems parameters*

We have noted that each subsystem and component will have certain input and output variables. The relationships between these inputs and outputs are the *parameters* of the subsystem. This is another term borrowed from mathematics, and refers to any fixed parts of a relationship. You will all be familiar with the relationship between two *variables* such as might be expressed by the equation

$$y = 10x + 15.$$

Here, y and x are the variables and 10, 15 and the $+$ and $=$ signs are all parameters of the relationship. No matter what values x and y may take, these parameters of the equation do not change.

parameter

A similar usage is possible for the relationships between input and output variables of a system component. For example, the relationship between the amount of light trapped by an autotroph and the amount of sugar produced forms one of the parameters of that component and can be defined for a particular *species* of autotroph in a particular *situation*. *Species* and *situation* are also parameters of that component.

2.3.7 *Systems response to input*

Assuming that all the previous questions have received satisfactory answers, the next stage usually involves applying some input to the whole system and measuring the changes induced in the output variables. For an ecosystem, however, this may be even more difficult than it is with, say, a social system.

We cannot simply go along to our system and increase the supply of energy to the system from the sun, for example. We may be able to cut down the energy supply to a small ecosystem, by putting some form of shading material over it; this at once alters the energy input, but may also change the rainfall, wind flow and a number of other variables.

An alternative approach, and one which has been widely used, is to apply some other test input, such as a fertilizer. This has given some interesting results but, if the ecosystem under examination is the sole example of its type, such an approach would lead to its complete destruction, and would clearly be highly undesirable. In addition, with large ecosystems, the cost of such experiments on the whole system may be prohibitive, since a large number of variables of a large number of components would need to be monitored. It is thus very difficult to conduct experiments in real life with whole ecosystems.

rare ecosystems

cost of experiments

SAQ 9

What systems technique could be used to overcome this problem?

This is where the greatest systems contribution comes to the study of ecology. You have already seen in Module 1 Deep-sea Container Ports how simulation can be applied in order to investigate the response of an expensive entity such as a container berth, or in Module 2 Air Traffic Control how it can be used, in a slightly different form, to avoid the risk of training operators with real systems. We shall now look at some techniques which we can use in a similar way with ecosystems.

3 Techniques for the modelling and simulation of ecosystems

3.1 General

We have seen that an important aspect of the function of ecosystems is the transfer of energy and materials between different organisms. This is broadly similar in many respects to the transfer of containers between different users, but has added complications due to the vast number of different components involved and the rapid changes in the number of these components which can occur.

SAQ 10

What term did we use earlier in referring to these changes in numbers of components?

SAQ 10

In order to cope with this greater complexity, we have to introduce some new modelling techniques which are particularly useful in modelling ecosystems, but can also be used for other types of system, as you will see in later modules.

A complete ecosystem model would be an extremely complex affair if it simulated all aspects of ecosystem behaviour. No one model has yet done this satisfactorily, but tremendous progress is currently being made, and more and more detailed models are being constructed.

Some simple examples will be found in the case-study section and a list of references is given in section 3.6; these give representatives of some of the most advanced models currently in existence, if you wish to pursue ecosystem modelling in greater depth.

All the models used are based on four main techniques. These are:

1 Compartment models (a form of flow or block diagram),
2 Difference equations,
3 Differential equations,
4 Matrix models.

You have already met some of these techniques. The model of the container berth and of the hydraulic level control system were formulated largely in terms of block diagrams and difference equations. Differential equations were also mentioned in connection with the level control system.

In the sections 3.2–5 we shall develop a particular form of block diagram – the compartment model – look in more detail at differential equations and examine the technique of matrix models.

3.2 Compartment models

Compartment models probably form the best basis for understanding all the other modelling techniques we have just mentioned. We can think of an ecosystem in terms of a set of compartments, between which are flows of energy and/or material. This has actually formed the conceptual basis of all the previous sections.

The fundamental unit of this model is the compartment or block, usually represented by a rectangle (Figure 8). Entering and leaving each block are flow paths, conventionally represented by solid lines, each carrying an arrow

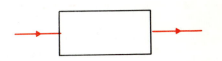

Figure 8   Compartment

to indicate the direction in which energy, sugar or some other commodity flows along that pathway.

This is a powerful means of summarizing a lot of information about a system; we used it in Figures 1 and 2 to illustrate the basic structure of an ecosystem. To give a quantitative description, we need to put numbers to the pathways and compartments, to indicate how much material (or energy) there is in each block and how fast it is moving along each pathway. The Massachusetts Institute of Technology (MIT) System Dynamics laboratory uses the term *level* for the amount present in a block, and *rate* for the speed of movement in each pathway. Returning to the level control system in Module 5, we can see the exact correspondence between the level of liquid, and the total amount present in the container or compartment. Similarly, we can see that there is a variable rate of outflow.

In the MIT System Dynamics convention, the rate of flow along a pathway is depicted by the symbol shown in Figure 9. Chemical engineers will recognize this as the symbol for a valve. The analogy is obvious, since a valve governs the rate of flow of water through a pipe.

There is a further symbol which is used in these flow charts. If the rates of flow are controlled by valves, something must control the degree of opening of these valves. This control is indicated by a dotted line on the flow chart linking the valve which is being opened to the factors which regulate its degree of opening. Thus, in a central-heating system, the degree of opening of a radiator valve might depend on the temperature of the room.

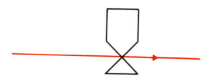

Figure 9   Flow symbol

Exercise

Can you draw a flow chart to indicate this?

Our attempt is shown in Figure 10.

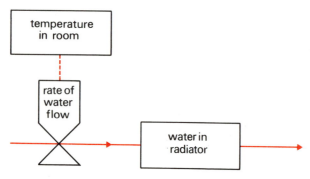

Figure 10   Flow chart of central-heating radiator

A further example might be taken from an ecosystem. Most plants need nitrogen from the soil in order to grow. The rate at which the plant takes this up from the soil probably depends both on the level of nitrogen in the soil and the level in the plant.

Exercise

Draw the flow chart for this nitrogen flow.

See Figure 11.

Figure 11   Flow model of nitrogen from soil to plant

A more complex example is given on p. 230 of the *Systems Behaviour* Reader. Have a look at this, concentrating on the physical flows indicated in the manner shown above.

3.2.1   *Control within ecosystems*

Modelling by this method introduces a very important theoretical concept. We have seen that the structure and functions of ecosystems can be summarized into three major components and five physical processes.

Exercise

List these components and processes.

See pp. 12 and 13.

We have not discussed the *regulation* of these processes. To *use* a compartment model of an ecosystem, we assume that the rate at which every process occurs depends entirely on the *rates* at which other processes are going on, and on the *levels* of other components (compartments) in the system.

regulation

These relationships may either be within each organism, or between organisms. For example, the rate of growth of a plant will depend on the rate at which it takes up energy from the sun and carbon dioxide from the air. The nature of these relationships form the parameters of the individuals and can often be demonstrated in simple experiments under controlled conditions. Interactions between organisms are more difficult to analyse, since by definition they can only be investigated in situations where several organisms are present, and hence the experimental situation is more complex than it is where only one organism is involved. The presence of one organism can only increase, decrease or have no effect on the rates of processes involving a second organism, but the *parameters* of these relationships between organisms are difficult to measure.

A few general concepts exist to cover such interactions: for example, a species is said to exhibit *density-dependent regulation* if its rate of growth or increase in numbers declines as more and more of that species live in a defined area. Similarly, if two species live in one area, and the rate of growth of one is depressed as the numbers of the other increase in that area, then the two species are in *competition*.

density-dependent regulation

competition

Exercise

Summarize a competitive relationship as a compartment model.

See Figure 12.

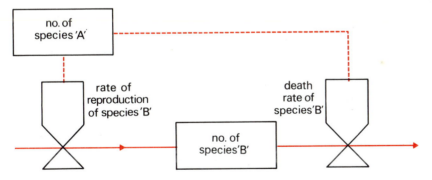

Figure 12   Compartment model of a competitive relationship

Note that, although we represent the interaction between the level of species A and the rate of growth of species B in a different way from the physical flow of growth, this interaction is the *result* of physical processes. For example, two plants may compete for light, so that if one absorbs some of the in-coming light, this is no longer available for the other. Similarly, in our central-heating example, the physical flow of heat to the person in the room causes him or her to apply a physical force to the radiator valve. All these physical processes could be represented by separate compartment/ physical flow models. However, it is much simpler to use the convention applied here to *summarize* such processes, although we should not lose sight of their existence when specifying the parameters of the various control relationships.

How do we put all this into numerical terms? In a compartment model, any change in a level in a compartment must be the result of inflows and out-flows to and from that compartment. Suppose, for example, that the compartment represented a bucket with a hole in it, which is being filled from a tap (Figure 13). The tap is running at a rate of $3\ dm^3$ per minute, and the water runs out of the hole at a rate of $2\ dm^3$ per minute.

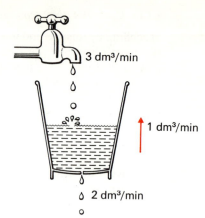

Figure 13 *Water flow from tap into leaking bucket*

See Figure 14.

Exercise

Summarize this example of the tap and the bucket as a compartment model.

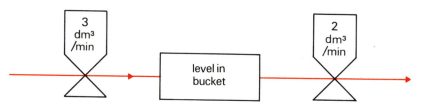

Figure 14 *Compartment model of tap and bucket*

Clearly the only *level* variable in this situation is the amount of water in the bucket. Thus the dynamic behaviour of this very simple system can be represented by the change in this level variable with time.

SAQ 11

How is the level of water in the bucket changing?

SAQ 11

This introduces the *dynamic behaviour* of the system as being represented by the changes with time of its *level* variables.

dynamic behaviour

Suppose a compartment receives an input at a *rate* of R units per minute and another input at a rate of F units per minute, while giving two outputs of O units per minute and P units per minute.

Can you write an equation which describes the way in which the *level* in that compartment changes?

The equation describing this situation is

Change in compartment level in units per minute
$$= R + F - (O + P).$$

Thus, every minute, the level in that compartment changes by $R + F - O - P$ units. This is a *difference equation* such as you met in Module 5.

Suppose the compartment has N inputs and Q outputs, all occurring at defined rates per minute, or per hour, or some other unit of time. The change in the level L in the compartment for each unit of time is often written ΔL (delta L, meaning change in L for some unit change in time).

Can you now write a more general equation for ΔL?

Our answer is

$$\Delta L = (\text{input}_1 + \text{input}_2 + \text{input}_3 + \cdots + \text{input}_N) -$$
$$- (\text{output}_1 + \text{output}_2 + \text{output}_3 + \cdots + \text{output}_Q)$$

or more concisely, writing each input as I_x, where x is a subscript, and outputs as O_y,

$$\Delta L = \sum_{x=1}^{N} I_x - \sum_{y=1}^{Q} O_y.$$

(If you are unsure about subscripts, refer to the *Systems Behaviour Mathematics Booklet*.) The sign $\sum$ means 'take the sum of'. In full,

$$\sum_{x=1}^{N} I_x$$

means 'take the sum of all the values of I from I_1 to I_N'.

This is the simplest possible form of a difference equation.

These are elaborated below.

3.3 Difference equations

Difference equations are so called because they refer to the *difference* between the values of some variable at two specified times; thus the change in a *level variable* between time now and a time, say, one minute later could be written as a difference equation and would give the change in level per minute.

In the example given earlier, all the terms on the right-hand side of the equation (all the rates of flow) were constant. In more complex situations, these flows might be *functions* of the rates and levels in other parts of the system. (If you are unsure of this meaning of function, refer to the Mathematics Booklet.)

Thus, we could quantify the model of nitrogen uptake into a plant given earlier on p. 26.

see Figure 11

We said there that the rate of uptake of nitrogen depended on the amount in the soil and the amount in the plant. Let us suppose we know that the amount taken up in one day is equal to the amount in the plant minus the amount in the soil, divided by a constant.

SAQ 12

SAQ 12

There are 20 units of nitrogen in the soil on 1 June, and 10 units in the plant. If the constant has a value of 10, can you write two difference equations which will describe the dynamic behaviour of the levels of nitrogen in soil and plant? Having done this, can you calculate how much nitrogen will be in each compartment on 10 June?

In fact, this model of the flow of nitrogen from soil to plant gives a very simple time-slicing simulation of the system, as defined in Module 1 for the container berth. After each slice of time, the variables take new values, and these are used to calculate the changes in the variables during the next time slice. This type of sequential operation is exactly how a digital computer works, calculating first one value, then the next, then the next, and so on, at a very high speed. This analogy with time-slicing simulation has led to the development of special computer languages which enable the computer to be programmed very easily to carry out such simulation exercises. An example of the use of such a language is given in the appendix to Spedding's *Grassland Ecology*, for those who are interested.

In addition to the suitability of difference equations for computer simulation, they are useful because many biological processes occur as discrete steps, such as the birth or death of an organism. When combined with the matrix techniques described later, they form a very powerful tool in the simulation of biological problems.

However, a number of biological processes are continuous and do not occur at discrete intervals of time. For example, an organism grows more or less steadily, not in sudden steps like birth or death. When a system which changes smoothly is simulated by a model employing difference equations, over a long period this can produce substantial discrepancies between the model and reality, because of cumulative errors in estimation. Let us look at how this happens.

Figure 15 shows a smoothly changing variable X, plotted against time T. We might wish to describe this approximately by a difference equation which stated that the change in X during a time interval from T_1 to T_2 was equal to the value of X at T_1 multiplied by a constant k.

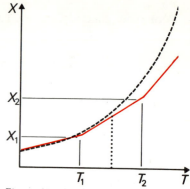

Figure 15 *Graph of smoothly changing variable X plotted against line T.*

SAQ 13

SAQ 13

Write out this difference equation.

The equation would be represented by the straight line joining the points X_1 and X_2 (see Mathematics Booklet). Now, what happens half-way between T_1 and T_2?

What is the value of X predicted by our equation?

The value of X midway between T_1 and T_2 is

$$X_1 + \tfrac{1}{2}\Delta X \qquad \text{or} \qquad X_1 + \tfrac{1}{2}kX_1.$$

Now, using the same criterion for the second half of the time period as we did for the whole period from T_1 to T_2, what should be the change in X during the second half of the period?

The change in X during the second half of the period should be

$\tfrac{1}{2}k \times$ (initial value of X for second half of time period),

i.e. $\qquad \tfrac{1}{2}k(X_1 + \tfrac{1}{2}\Delta X) \qquad \text{or} \qquad \tfrac{1}{2}k(X_1 + \tfrac{1}{2}kX_1).$

The value of X at the end of the time period (T_2) should be the value at the midpoint plus this revised change in the second half.

What would this value of X be?

Calculated in this way, the value of X at T_2 is

$$(X_1 + \tfrac{1}{2}kX_1) + \tfrac{1}{2}k(X_1 + \tfrac{1}{2}kX_1) = X_1 + \tfrac{1}{2}kX_1 + \tfrac{1}{2}kX_1 + \tfrac{1}{4}k^2X_1$$
$$= X_1 + kX_1 + \tfrac{1}{4}k^2X_1.$$

This is clearly greater than the original prediction.

If the original prediction is then used for succeeding time periods, this error will build up and become greater for each succeeding period. We could overcome this partly by deliberately overestimating the constant k, but this would not be very satisfactory unless we knew exactly by *how much* to overestimate it.

How can we overcome this problem? One way would be to recalculate the variable (X in the example above) at extremely short intervals of time, but this would become very tedious. Ideally, we want to find a way of calculating the value of the variable at *any* instant of time.

Exercise

Do you know how this can be done? (Hint: Look at the Mathematics Booklet.)

The answer is to use, instead of a difference equation, a differential equation. In many cases this can be used to relate the value of the variable to *any* instant in time, not only to multiples of the chosen difference interval.

3.4 Differential equations

Before reading this section, remind yourselves of the symbols and terminology of calculus, as given in the Mathematics Booklet.

Consider a plant growing in a completely open space well supplied with all soil nutrients. The sun's light energy falls uniformly and at a constant rate onto all the space, and the plant's growth rate depends solely on how much of this energy it can trap. Clearly, the greater area the plant occupies, the more energy will be trapped before it falls on the soil and is wasted. Now, the area the plant occupies is roughly proportional to its size, and its size will depend on how fast it has been growing and for how long. This situation can be formulated in terms of a *differential equation*. Representing the plant's size by w and the time for which it has been growing by t, then, in the Leibniz notation (see the Mathematics Booklet), its rate of growth at that time is

$$\frac{dw}{dt} \quad \text{or} \quad \frac{\text{Change in size}}{\text{Change in time}}.$$

We have already stated that rate of growth is proportional to amount of energy trapped, which is proportional to size w. In mathematical shorthand this is represented by

$$\frac{dw}{dt} \propto w.$$

This can be rewritten

$$\frac{dw}{dt} = kw, \tag{1}$$

where k is an arbitrary constant.

What does this tell us? It says that the rate of increase in size with time, is equal to a constant multiplied by current size. In looking at the dynamic behaviour of a system, we want to know the value of a variable or variables (such as the size of the plant) at any instant in time. Can we find this from equation (1), which is a differential equation relating the differential (or rate of change) of the unknown variable with respect to time, to the value of the unknown variable? We know how fast it changes when it is a given size, but we do not yet know what size it will be at a given time. In the notation above, we know the relationship between w and $\frac{dw}{dt}$. We require to know the relationship between w and t.

In the Mathematics Booklet, we saw that $\frac{dw}{dt}$ can be written as $f'(t)$. We also saw that there is another function, written $f(t)$. This is the same as w. Put another way, if

$$\frac{dw}{dt} = f'(t),$$

then $w = f(t)$. $\tag{2}$

This relationship (2) is the one we require. By definition, we call f the *integral* of f'. We already know the form of f' from equation (1).

integral

SAQ 14

What is the form of the function f'?

SAQ 14

Now, ideally, we should like to integrate this function f', using the rules given in the Mathematics Booklet.

integration

SAQ 15

Can you see why it is difficult to integrate the function f'?

SAQ 15

31

Thus, we have to rearrange equation (1) into some form where we can integrate it, and find the unknown w. This is called *solving* the differential equation and in this case can be done by a technique known as the *separation of variables*.

solving

separation of variables

It involves rearranging equation (1) so that terms including t are all on one side, and those including w on the other. To do this, we have to pretend that dw and dt are quantities which can be manipulated as if they were numbers, as if $\dfrac{dw}{dt}$ was equivalent say, to $\dfrac{3}{2}$ or $\dfrac{x}{y}$. This is not strictly true, but is a valid and convenient assumption in this case. Knowing when such manipulations are justified is part of the art of solving differential equations.

Applying this to equation (1),

$$\frac{dw}{w} = k\, dt$$

or $\quad \dfrac{1}{w} \times dw = k\, dt.$

The two variables are now *separate* and both sides can be integrated, the left with respect to dw and the right with respect to dt.

Can you do this for the right-hand side?

$$\int k\, dt = k \int dt,$$

since the integral of (a constant multiplied by a function) is equal to the constant multiplied by the integral of the function. (A helpful analogy may be to regard the integral sign as meaning the summation of all the 'little bits' of function times differential. If each of these bits contains the same constant, this constant can be taken outside the summation and multiplied once by the final result.)

Since $\int dt = t$, the right-hand side becomes $kt + c_1$ (remember the constant of integration – see the Mathematics Booklet).

On the left-hand side,

$$\int \frac{1}{w}\, dw = (\log_e w) + c_2$$

(remember $\int \dfrac{1}{x}\, dx = \log_e x$ from the Mathematics Booklet).

Completing the expression,

$$(\log_e w) + c_2 = kt + c_1.$$

We can combine c_2 and c_1 into a single constant c,

i.e. $\quad \log_e w = kt + c.$

Now it is convenient to put c equal to $\log_e w_0$, where w_0 is the size of the organism at zero time, that is, at the beginning of the period of interest.

Thus

$$\log_e w = kt + \log_e w_0.$$

One property of logarithms is that

$$\log_e w - \log_e w_0 = \log_e \frac{w}{w_0};$$

using this we obtain

$$\log_e \frac{w}{w_0} = kt.$$

Another property of logarithms is that, if $\log_a b = x$, then $a^x = b$.

So, if $\quad \log_e \dfrac{w}{w_0} = kt,$

then $\quad e^{kt} = \dfrac{w}{w_0}.$

Using this property, we obtain the relationship

$$\frac{w}{w_0} = e^{kt}$$

or $\qquad w = w_0 e^{kt}.$ (3)

This, then, is the function relating w and t, the desired *solution* of the differential equation (1). **solution**

Solving differential equations demands some degree of mathematical expertise. If you wish to know more, you may like to look at Chaston's *Mathematics for Ecologists*,* chapter 8. An alternative approach is to use an analogue computer, such as the one supplied in your Home Experiment Kit. Home Experiment Book 2 contains a series of problems for you to solve using the kit. In addition, you will see the use of an analogue computer to solve a biological problem in the television programme accompanying this module.

Differential equations do not always contain just the first derivative of a function. They can contain higher derivatives of any *order* (see the Mathematics Booklet), and are classified according to the highest derivative involved. Thus an equation with a third derivative **order of differential equation**

$$\frac{d^3 y}{dx^3} \quad \text{or} \quad f'''(x)$$

would be a *third-order differential equation*.

Where the dependent variable in a differential equation occurs alone, it is called a *linear equation*. If, however, it contains terms such as (dependent variable)^2, or other functions of the dependent variable, it is a *non-linear equation*. Non-linear equations are much less tractable mathematically and can usually only be solved approximately or with the aid of a computer. Unfortunately, most biological models seem to involve non-linear equations. **linear equation**
non-linear equations

It is difficult to know which equation type is better suited to ecological problems: difference equations are easier for a non-mathematician to comprehend, and are easily programmed onto a digital computer. However, the theory of differential equations is very well developed, so that if a system *can* be modelled in terms of a series of differential equations, a mathematician can often predict many of its behaviour patterns without the need explicitly to solve the equations. This will be covered more fully in Module 7 The Human Respiratory System.

As we noted earlier, many biological processes, such as growth, occur continuously and so are more accurately represented by differential equations than by the discrete changes demanded by a model using difference equations. However, the *measurement* of growth of plants or animals usually only takes place at intervals, and their growth is usually calculated as a difference between successive weights. For this situation, the difference equation is quite an adequate representation, and also for population problems where discrete events – births and deaths – are involved. **measurement**

A particular population model, using difference equations allied to matrix manipulations, is described in the next section.

*Chaston, I. (1971) Mathematics for Ecologists, *Butterworths*.

3.5 Transition matrices

In its simplest form, the technique of transition matrices is a compact method of representing changes in a population of animals. It has been extended to describe other situations and has the great advantage that, although it is based on a difference-equation approach, it uses quite powerful mathematical techniques evolved for the solution of matrix problems. This gives many of the theoretical advantages of differential equations, combined with the intuitive simplicity of the difference equation.

In its simplest form, the *Leslie matrix* comprises firstly a column of numbers, called a vector, which is used to represent the age structure of a population of animals. Thus, suppose an animal, such as the *Daphnia* used in your home experiment, lives for three days (*Daphnia* actually live much longer than this). At any time, in a population of this organism, there would be animals which were 1 day, 2 days and 3 days old. The number of animals in each age group describes the *age structure* of that population. Thus, the age structure of this population might be represented by a column vector as below:

Leslie matrix

$$\begin{bmatrix} 5 \\ 10 \\ 9 \end{bmatrix} \quad \begin{array}{l} \text{animals 0–1 day old,} \\ \text{animals 1–2 days old,} \\ \text{animals 2–3 days old.} \end{array}$$

The brackets are the conventional representation of a vector, and imply that these numbers are all similar measures, occurring in a set order.

Now, if this animal reproduces asexually, as does *Daphnia*, the animals in each age class will reproduce during the time interval from t_0, when the population was described by the previous vector, to time t_1. The rate at which they reproduce is called their *fecundity*. It is measured by the average number of young born, in the interval from t_0 to t_1, to an organism of given age at t_0. On average, an animal aged from 0 to 1 day may produce 2 offspring per day, an animal aged from 1 day to 2 days may produce 4 offspring per day, an animal aged from 2 to 3 days may produce 5 offspring per day and so on. Another vector, this time a row vector, can be written to represent these age-specific fecundities, thus:

fecundity

$$\begin{bmatrix} 2 & 4 & 5 \end{bmatrix} \quad \text{young born to}$$
$$\text{0–1} \quad \text{1–2} \quad \text{2–3} \quad \text{day-old animals.}$$

Similarly, in the time from t_0 to t_1, some of the animals will die. Those that do not will, at t_1, enter the next oldest age class. Thus, at the end of one day, animals aged between 0 and 1 day at the beginning of that day, will now be aged between 1 day and 2 days. Suppose 20 per cent of the animals aged between 0 and 1 at t_0 have died by t_1, one day later. Then 80 per cent of the original animals will have made the *transition* from age class 0–1 to age class 1–2. This transition from one age class to the next is comparable to a movement from one place to another.

transition

SAQ 16

SAQ 16

If a ball is held at arm's length and then released, repeatedly, in what proportion of the occasions will it make the transition from arm to ground? In what proportion of occasions will it make the transition from arm to ceiling?

The proportion of the number of occasions on which a given transition occurs is a measure of the *probability* of that transition (see Module 1 for a discussion of probability). Any such transition from one place to another will have a definable probability of occurring during a given time interval. Thus, the probability of a man moving from his home to his place of work on any day will be about $\frac{5}{7}$, assuming a five-day week (and no holidays!). The probability of his making the reverse transition, from work to home, should also be $\frac{5}{7}$ for any day.

probability

Similarly, any *change of state* will have a definable probability of occurring in a given time interval. A man may change his job, which is just one possible change in his state. In any year the probability of his changing his job from, say, shepherd to dairyman might be 0.05 implying that one in twenty of all shepherds is likely to become a dairyman in any one year. Likewise, for a man aged twenty-eight, there will be a given probability of his age (his *state*) changing from twenty-eight to twenty-nine in one year. This will be considerably greater than 0.05.

change of state

Returning to our example of the age transition in *Daphnia*, similar probabilities will exist for each age class, and this situation can be summarized by a *transition matrix* as shown below. Each transition is expressed as the probability, equivalently the proportion, of animals in age group i surviving to age group $i + 1$ during one time interval. This number is placed in column i and row $i + 1$ of the transition matrix. Thus suppose the probability of survival for organisms in our original population into the next oldest age class is:

From age 0–1 to age 1–2:   0.8;
From age 1–2 to age 2–3:   0.8;
From age 2–3 to age 3–4:   0.0   (lifespan is 3 days).

We summarize this in our matrix as follows:

Age now	0–1	1–2	2–3
Age one day later			
0–1	0	0	0
1–2	0.8	0	0
2–3	0	0.8	0
3–4	0	0	0

Each element in this matrix now represents the probability of transition from an age group represented by the column number to an age group represented by the row number, during one time interval.

(*Note.* If you are familiar with matrix operations, you will notice that the order of priority is reversed here; normally matrices are considered in the order row – column. Here the priority is column – row. This does *not* affect the rules of multiplication etc.)

SAQ 17

SAQ 17

What do the zeros in the matrix mean?

In this matrix, we have all zeros in row 1, suggesting, reasonably, that no animal can age backwards into this row. However in practice this can apparently occur. Can you think how this is?

The answer is, of course, the birth of new organisms, so that each animal in a given age class introduces a number of young into age class 0–1 in each time interval. This number will be equal to its fecundity, and we incorporate this into our matrix by substituting the row vector for fecundity given earlier into the first row of the matrix, thus:

$$\begin{bmatrix} 2 & 4 & 5 \\ 0.8 & 0 & 0 \\ 0 & 0.8 & 0 \end{bmatrix}.$$

At the same time, for convenience, we have removed the bottom row of zeros entirely, because it represents non-existent members of the population (since all die at an age of 3 days there can be no 3–4-day-olds). This gives us a square matrix where all the *possible* changes in state are represented, using the convention that a change in state from state i to state j is found in the matrix in column i row j.

Can you construct a transition matrix for the man going to and from work, referred to earlier?

(Hint. Remember he stays at home sometimes.)

What use is this? We have defined that if there are five animals 1–2 days old at $t = 0$, then during day 0–1, on average, one will die by the end of the day and each one present initially will give birth to four young. Thus, at $t = 1$ day there will be four adults which have survived to age 2–3 days and $5 \times 4 = 20$ juveniles (0–1 days old) born to the five adults who were 1–2 days old during day 0–1. Similarly members of each age class will have given birth to the appropriate number, and a proportion will have survived to the next age class. As a result there will be a new vector, describing the new age structure of the population. Each number in this vector for the second day will represent the total number of animals which have made the transition *into* that age class, either by birth or natural ageing.

This total is equal to the sum of (the probabilities of transition from other age classes into that age class, multiplied by the number in each of these other age classes during the preceding time interval).

In mathematical notation this is

$$n_i(t + 1) = \sum_{j=1}^{k} P_{ji} \times n_j(t), \tag{4}$$

where $n_i(t + 1)$ is the number in age class i at time $t + 1$, $n_j(t)$ is the number in any age class j at time t and P_{ji} is the probability of transition from class j to class i. In our matrix P_{ji} is represented by the figure in row i column j. Now the rules for the multiplication of matrices state that the result of multiplying a matrix by a column vector is a new vector, with the same number of elements as the original. The element in row i of this new vector is equal to the sum of (the elements in the original vector times the corresponding elements in the ith *row* of the matrix). Thus, for the population vector and the matrix given earlier, we have

$$\begin{bmatrix} 2 & 4 & 5 \\ 0.8 & 0 & 0 \\ 0 & 0.8 & 0 \end{bmatrix} \begin{bmatrix} 5 \\ 10 \\ 9 \end{bmatrix} = \begin{bmatrix} 5 \times 2 + & 10 \times 4 + & 9 \times 5 \\ 5 \times 0.8 + & 10 \times 0 + & 9 \times 0 \\ 5 \times 0 + & 10 + 0.8 + & 9 \times 0 \end{bmatrix} = \begin{bmatrix} 95 \\ 4 \\ 8 \end{bmatrix}.$$

Examination of the elements in the new vector will show that each element corresponds exactly with the summation (4) given earlier. Each element gives the sum of all the possible transitions into the age class represented by that element. Hence we have the new vector describing the population after one day (t_1), obtained by taking

[transition matrix] $\times$ [vector of population at t_0].

Similarly, after a further interval of a day, at t_2 the population is

[transition matrix] $\times$ [vector of population at t_1]

or   [transition matrix]^2 $\times$ [vector of population at t_0].

In general, representing the population vector at a time n intervals from the start (t_n) by V_n, the vector at t_0 by V_0 and the transition matrix by M, we have

$$V_n = M^n V_0. \tag{5}$$

(The bold letters are conventional representations of matrices and vectors.)

This is the general formula for transition matrices. You should now attempt to calculate the population vectors for one or two further days, following the example given earlier. For longhand multiplication it is easier to multiply the matrix by each new vector as in (4) rather than using the formula (5) (although this gives greater opportunity for error).

On your computer terminal, there is a program which will repeat this type of multiplication for you. You are recommended to use this to carry out the above exercise up to ten times. You will notice that after five or six multiplications the ratios of the numbers of animals in each age class to the number in the youngest age class tend to have the same values after each multiplication, although the actual numbers are still increasing. When these ratios reach steady values it indicates that the age structure of the population has become *stable*.

stable age structure

This may or may not be realistic. However, at this point, the population does exhibit the interesting feature that it has a rate of increase in total numbers which is proportional to the total number present. This is similar to the growth form proposed earlier for a plant, where, in terms of a differential equation,

$$\frac{dw}{dt} = kw.$$

Now, interestingly, the constant corresponding to this k for the matrix model can be derived mathematically from the transition matrix and, as with the theory of differential equations, so can certain other properties such as the inherent stability of the population structure. The whole model can be extended to cover sexual reproduction, and to the situation where survival and fecundity are not simple numbers, but functions of the numbers in that or other age classes. It can be further generalized to cover energy flows and leads to a number of very important conclusions which are regrettably beyond the scope of our discussion.

Anyone who is interested is recommended to read Usher's paper* in *Mathematical Models in Ecology*.

SAQ 19

SAQ 19

What types of model might be most suited to the following studies?

(a) A population of deer, who start to breed at an age of 2 years and have a maximum lifespan of 10 years.

(b) A population of bacteria, reproducing continually.

(c) A system of grassland, cattle and lions.

3.6 Further reading

If you feel you would like to know more about the use of models in ecology, the publications listed below give some advanced examples.

1 Jeffers, J. N. R. (ed.) (1972) *Mathematical Models in Ecology* 12th Symposium of the British Ecological Society, Blackwell. Papers in this volume by G. R. Conway and G. Murdie, D. W. Goodall, C. Milner, and M. B. Usher.

2 Jones, J. G. W. (ed.) (1970) *The Use of Models in Agricultural and Biological Research*, Grassland Research Institute. Papers by N. R. Brockington, and J. G. W. Jones.

3 May, P. F., Till, A. R., and Cumming, M. J. (1972) 'Systems analysis of [35]sulphur kinetics in pastures grazed by sheep' *Journal of applied Ecology*, vol. 9, no. 1, pp. 25–49.

4 Ross, P. J., Henzell, E. F., and Ross, D. R. (1972) 'Effects of nitrogen and light in grass–legume pastures – a systems analysis approach' *Journal of applied Ecology*, vol. 9, no. 2, pp. 535–56.

5 de Wit, C. T. Brouwer, R., and Penning de Vries, F. W. T. (1971) 'A dynamic model of plant and crop growth' in Cooper, J. P., and Wareing, P. F. (eds.) *Potential Crop Production*, Heinemann Educational.

You will find that some of the terminology in these may be unfamiliar. This does not affect the principles employed, which should be common to all systems studies.

* Usher, M. B. (1972) 'Developments in the Leslie matrix model' in Jeffers, J. N. R. (ed.) Mathematical Models in Ecology 12th Symposium of the British Ecological Society, Blackwell.

4 Testing ecosystem models

In Dale's article, the final phase of a systems study is called the analysis phase (*Systems Behaviour* Reader, p. 255). It involves the *validation* of the model that has been constructed. We have seen some of the techniques which might be used to construct a model of an ecosystem and how such a model, when quantified, can be used to *simulate* the behaviour of the real ecosystem. We may wish to do this simply to demonstrate our ideas of the ecosystem's structure or alternatively to experiment with the model, rather than with the real system, as suggested on p. 24. In either case, we shall need to know how accurately the behaviour of our model mimics the behaviour of the real system. This section deals with the problems involved in testing this.

No model of a real system can ever reproduce exactly the behaviour of all possible variables within the system. Some of the subsystems may be only modelled approximately and others may not even be recognized. Thus the correspondence between the values of variables in the real system and those predicted by the model is unlikely to be exact. We might decide to draw graphs of the values of the variable from the model and from the real system as they change with time and then compare the shapes of these graphs. If they appear similar, we might say that we are satisfied with our model. However, you will have learnt from Module 3 Industrial Social Systems that this is not very satisfactory.

SAQ 20

Why is it an inadequate test of the model to say that the variable values appear to be similar?

In Module 3 you were introduced to two *statistical tests*, which enabled you to decide whether a particular set of results could have arisen by chance alone, or whether there was likely to be some *reason* for such results. In *testing* a model, we use these techniques and some more advanced ones. Here, however, we are more concerned with the general principles of model testing than with any particular techniques.

All the statistical tests which are in common use are based on the concept of *probability*.

SAQ 21

Out of the last 100 days that I have travelled to work, I have arrived late on 23 occasions. On this information alone, what is the probability that I shall arrive late on my next journey to work?

In general, if we obtain a set of values for our variable from the real world and a set of predicted values from the model, then our statistical tests will tell us the probability of these sets of values coinciding *by chance alone*. If this probability is low, then we can be fairly happy that there is some real relationship between our model and the real system. This is stated more formally in terms of what is called the *null hypothesis*.

The null hypothesis states that the values of the variable from the model and from the real system are completely unrelated, and any apparent relationship exists by chance alone. The null hypothesis can then be *rejected* if our statistical tests show that there is only a very small probability of the

validation

simulate

SAQ 20

statistical tests

model testing

probability

SAQ 21

null hypothesis

actual sets of results occurring together by chance. To do this, we usually need to estimate what is the probability of occurrence of the particular set of results from the real world, and another probability for the results from the model, in order to estimate the probability of their matching together.

SAQ 22

> How might we estimate the probability of obtaining a given set of values for a variable from the real world?

The obvious answer, of repeated measurement, may be fine for some systems. Unfortunately, as we saw earlier, ecosystems tend to have many variables which should all be measured in order to give a complete picture of its behaviour. This is likely to prove time-consuming and expensive. Similarly, the main input to the system, solar energy, is not constant, but varies both on a fairly regular seasonal basis and on a more irregular daily basis as cloud cover changes. Along with this go other changes in temperature, wind speed, and so on, all of which may affect the system's functioning.

One solution might be to observe a large number of similar systems, over the same time period. This is called *replication*. There are two drawbacks to this: one is that many ecosystems are unique, and so there may be only one to observe; a second is that we cannot be sure that all conditions are the same on all the sites observed. We might, for example, want to look at oak woodlands as ecosystems. These occur on a number of different types of soil, on level ground, on slopes, on slopes facing north, south, east, west, and in a variety of other places. It is questionable whether we can regard these as similar systems.

replication

What about the probability of obtaining a given set of values from the model? Again, we could run the models a number of times, and record the results. However, all the models described in this module are basically *deterministic*. This means that, for a given input, they will always give the same set of values for the output variables at a given time. For example, consider the model of a growing plant in section 3.4. The equation relating its size to time was

deterministic

$$w = w_0 e^{kt}.$$

For any value of w_0, the value of w at a given time t is determined solely by e and k. These are both constants, so, *every time such a model is run*, it will produce the same set of values. A similar problem occurred with the simulation of the container berth in Module 1. There, to overcome this problem, the input to the model – the number of containers – was varied at random, according to some defined scheme.

> How could this be applied to the model of the growing plant?

The value of w_0 could be chosen at random, using a scheme, like that for the simulation of the container berth, of sampling from a distribution. This would give the model a *probabilistic* input, although it would still be a *deterministic* model. Another approach might be to conduct a number of simulation runs using different values for the parameters of the model.

probabilistic

SAQ 23

> Which parameter of the plant growth model could be varied?

Here again we come up against a problem. We have to know what range of values of the parameters to use, and what the distribution of values should be. In the simulation of the container berth, the distribution of ship arrivals and loads was assumed to be known. How, in an ecosystem, could we find out such details? With the model of the growing plant, we could measure the value of k in a number of experiments and use this in our simulation. We should then have to test our simulation results against *another* series of experiments, or our test would be meaningless.

SAQ 24

Why could we not use the same set of experiments?

SAQ 24

Once again we come up against the difficulty of conducting experiments on whole ecosystems.

SAQ 25

List some of the problems of conducting experiments on whole ecosystems.

SAQ 25

Thus it is difficult to conduct separate series of experiments on the system. Some people have derived parameter values for ecosystem models from a single ecosystem, and then tested the model against the behaviour of the system at a later date. Even this is questionable, as it contains an element of testing the system against itself.

Ideally, parameter values for a model of an ecosystem should be obtained from experiments on the subsystems of the whole system, preferably in the laboratory, and then these values should be used in models tested against the whole system.

This gives you some idea of the problems involved in testing ecosystem models. It is a difficult task, but every model must be tested, though the temptation to use an apparently satisfactory model with only a minimum of rigorous testing is hard to resist.

5 Grassland sheep production

5.1 Introduction

So far, we have considered in fairly abstract terms what we mean by ecosystems and we have looked at some of the techniques we can use to model them. We are now going to compare this outline with an example from the real world by undertaking a systems study of sheep production. More specifically, we are going to look at the system whereby mutton and wool are produced from grass and other feedingstuff within the United Kingdom. We shall use the general framework for a systems study given on p. 19, using the specific techniques developed in section 3 where they are appropriate to set up fairly detailed models of parts of the system. You should also bear in mind the techniques and approaches you have learnt from earlier modules. Try to see if these can be applied to this particular case study, as they are not applicable just to the particular case study with which they are immediately associated. By this means, you should get a more complete picture of each case study than could be obtained using only one or two techniques.

SAQ 26

What is the first question we ask in systems study?

SAQ 26

5.2 Reason for studying the system

Our primary reason for studying sheep production is to illustrate general material which has come in the earlier part of this module. However, we also want to understand *how* the system functions, and to determine what affects the production of meat and wool in this system, since most of us are consumers of these products. This academic interest allows us to take the broadest possible view of the system, since we are concerned only with understanding the system and not specifically with *managing* it.

Since the module is concerned particularly with ecosystems, we should at this point ask whether or not the system we are going to study *is* an ecosystem. We can answer this with reference to Miller's definition in section 1.1.

SAQ 27

Is the system of sheep production an open system?

SAQ 27

SAQ 28

Can you name some of the plants and animals, organic residues, etc. that this system contains?

SAQ 28

SAQ 29

Does the system involve a flow of (a) energy and (b) material among these components?

SAQ 29

Clearly, from the definition given, it is an ecosystem. We also know that it has one feature not listed in this definition, in that most sheep are owned and managed by humans. This is not always a feature of ecosystems and it indicates that, like many other systems, this system is a part of some larger system or systems. We shall consider this in more detail in relation to our second systems question.

Exercise

What is the next question to ask?

See list in section 2.3.1.

5.3 Systems boundaries

As an ecosystem, grassland sheep production is that part of the world ecosystem or biosphere which specifically involves sheep and grass in the United Kingdom. It therefore embraces all the land on which grass grows and is eaten by sheep, the sheep themselves, their lambs, internal parasites, the animals living in the soil and so on.

Sheep are produced for sale either to other farmers or for meat – an economic transaction – so that the system in which we are interested is also part of the economic system which we know as British agriculture. This is a part of the British economy and hence of the world economic system. This is illustrated in Figure 16.

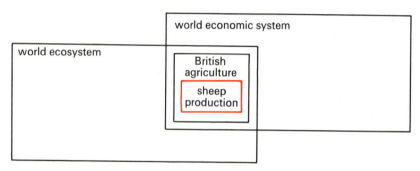

Figure 16   Systems map showing the position of sheep production in Britain as a subsystem of British agriculture, which in turn is a subsystem of the world ecosystem and the world economic system

Within our system we must also include the men who own and care for the sheep. They will be responsible for selling the sheep to butchers or wool processors. For convenience, we are going to consider that the butchers and wool processors are outside the boundary of our system. Similarly, the shepherd may find it necessary to purchase grass or other feedingstuff from merchants or other farmers. Again, for convenience, we are excluding these two from our system.

As a general criterion, anything which affects our system but is not affected materially by it is deemed to be outside the system's boundary. Thus, the atmospheric system and the hydrologic system (global water system) both affect sheep production, but the converse effect is negligible. They therefore form part of the *system's environment*. The boundary of the system in which we are interested is shown diagrammatically in Figure 17.   **systems environment**

Note that man, as a controller, has a central position in our system. The men involved are the farmers, their families and their workers. These people could be considered as a separate socio-technical subsystem, like those studied in Module 3.

> As an aside, we might speculate that, in practice, man is involved in almost every part of the global ecosystem, although his influence in many parts may be minimal. To speak of *natural* ecosystems, implying that these are unique in being unsullied by the hand of man is probably a nonsense. There is a continuous range from very closely managed ecosystems, like greenhouses, to those, such as tropical rain forests, where man has little influence, even today.   **'natural' ecosystem**

SAQ 30   **SAQ 30**

Because sheep production is closely controlled by man and is part of an economic system, ought we to modify the list of systems questions in section 2.3.1?

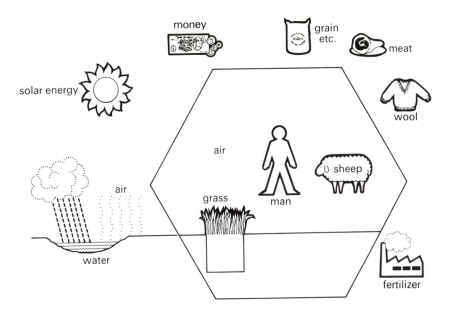

Figure 17   Systems components and boundaries of the system of sheep production

5.4   Purpose of the system

In practice, the sheep production system has three *goals* set by us. The first goal is a biological one, the production of meat and wool for consumption by people outside the system.

systems goals

The second goal is an economic one, to provide a livelihood and profit for the farmer involved. This profit may not come directly from the sheep: in some farming systems it may be necessary to grow grass for a period on any field in order to repair or build up the structure of the soil in between periods of intensive cropping, for example, with wheat or barley. In such a case, the sheep are used solely as a means of harvesting the grass and producing dung, which is incorporated into the soil. More detail as to how this occurs will be given in section 5.6.4. There may be little or no direct profit from the sheep, but an increased profit may accrue to the following crop due to this improvement in soil structure. This again illustrates the important general point that you should look at all aspects of a system when evaluating its effectiveness in economic or other terms, as in Module 5.

The third goal of the system, in this case rather an ill-defined one, is to provide a job and 'way of life' for the farmer and his family, and to maintain a landscape in which others may enjoy their recreation. Goals of this type were discussed in Module 3 in terms of 'job satisfaction'.

It is important to contrast these three specific goals with those of totally unmanaged ecosystems. As we discussed earlier (section 2.3.2), these have no goal, although the individual species involved each may have the goal of survival, usually expressed in terms of the maximum possible rates of growth and reproduction. Part of the problem in managing ecosystems for economic purposes, as in farming, lies in reconciling not only the three goals given earlier, but also the goals of the individual organisms.

5.5   Components and subsystems

Our next systems question demands that we specify the subsystems within the whole system. We have already specified four major components.

SAQ 31

SAQ 31

What were these major components of the sheep-production system?

In addition to these, there is a fifth component, a food store, whose function will become apparent later in the module (Figure 18).

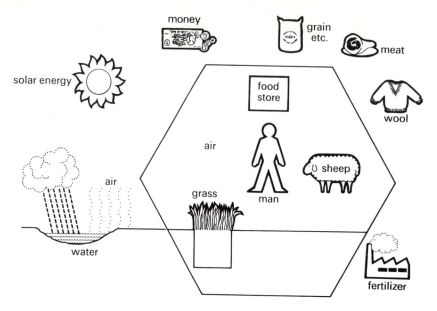

Figure 18   Systems components and boundaries, including food store

In section 1.3 we looked at the flows of material and energy between the components of generalized ecosystems, setting up a very simple *compartment model* of the system (Figure 1). We now want to do the same for the sheep-production system.

First we need to establish the correspondence between the compartments of the system of sheep production and the components of the generalized system.

SAQ 32

SAQ 32

Which are the autotrophs, heterotrophs and non-living parts of the system of sheep production?

We can now fill in the material and energy flows between these components of the system. This is done in Figure 19.

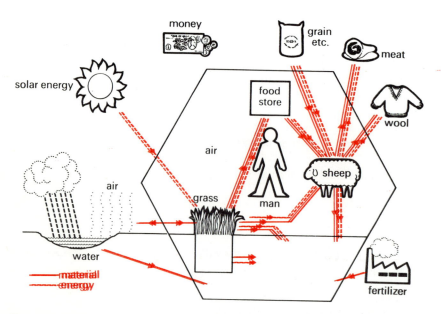

Figure 19   Material and energy flows in the system

The soil is shown here as a separate subsystem; it could almost be regarded as another ecosystem in its own right, since it harbours a large number of living organisms which carry on processes of material and energy transfer. We shall look at this subsystem in more detail in the next section, where we consider the subsystems to determine their variables and parameters.

5.5.1 *The position of man in the system*

Note that in Figure 19 man does not receive any material or energy flows. Remember that we specified man as the 'controller' of the system, and not as the consumer of meat or wool. In this context he could be likened to the air traffic controllers studied earlier. In the sheep system his role is to collect *information* about the functioning of the system and to make use of this information in controlling its behaviour. In Figure 20 we illustrate this in the system diagram by a series of dotted lines, as in the *control* of the radiator valve, Figure 10. Compare this with the discussion of information processing on p. 66 of Module 2. Note that man in this system can *store* information about the system, and base decisions on this stored information. This contrasts with the regulation of an unmanaged ecosystem, which arose solely from the physical interactions in the system, and where there was no attempt at storage of information for later control decisions.

information storage

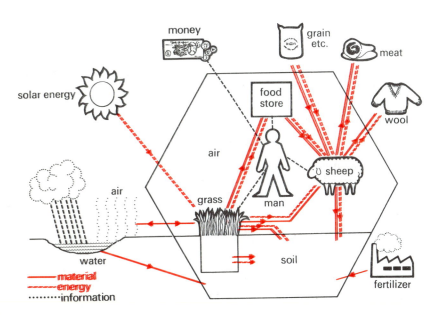

Figure 20   Information flows and control of the system

While viewing the television programme associated with this case study, try to analyse the inputs of information to the man who is controller of the sheep system, and try to see what decisions he makes on the basis of information received. Try to compare his situation with that of the air traffic controller.

Many of the decisions which a farm manager has to make are financial ones. This is indicated in Figure 20 by the flow of control information between man in the system and the economic system outside its boundary.

5.6   Variables and parameters of the subsystems

5.6.1   *Grass*

The flowering plants are grouped into two classes – the Dicotyledons and the Monocotyledons. Grasses belong to the Monocotyledons, along with plants such as lilies, onions, leeks and rushes. A typical grass plant, with the names

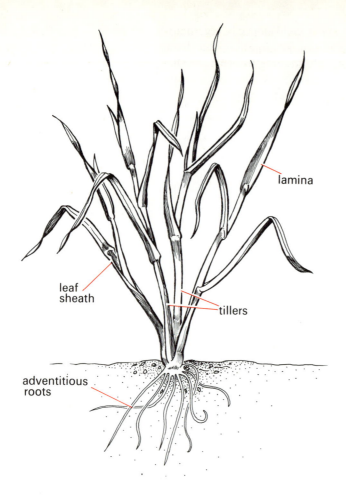

Figure 21   A typical grass plant

of its parts, is shown in Figure 21. At first sight, there might appear to be little
to distinguish grasses from plants like nettles or wall-flowers, which we do
not *normally* find being grazed by sheep. A grass plant consists of strap-like
leaves, *laminae*, borne at intervals up a vertical 'stem', called the *leaf sheath*.
However, if we examine a grass plant carefully we find that what looks like a
stem is actually a tubular continuation of the leaf, and each leaf can be
stripped away to reveal a series of such concentric tubes. If this is continued
for long enough, the true *stem* from which all these leaf sheaths grow will be
found to occupy only a few millimetres at the base of the plant. A much
enlarged section through such a true stem is shown in Figure 22.

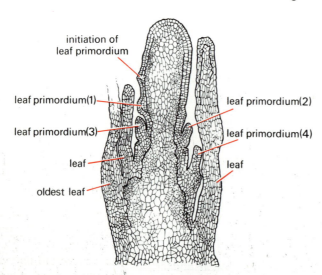

Figure 22   Vertical section through the true stem of a grass plant (highly magnified)

At this stage we suggest you get the hand lens from your Home Experiment Kit, and try to find this tiny stem in a grass plant taken from your lawn or a convenient field. Alternatively, you could look at a leek, which has a similar structure.

When plants grow, they do not increase uniformly all over. Growth actually takes place in a few specialized sites called *meristems*. These are usually found around the tips of stems, where the youngest leaves are found and in the *axils* (sort of inverted armpits), where the leaves join the stems. Most of the older parts of the plant are not capable of growing under normal conditions. The positions of the meristems in a grass and in a nettle, which is not normally grazed, are indicated in Figure 23.

meristem
axil

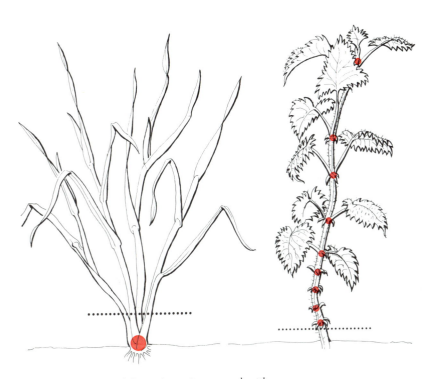

Figure 23   Positions of the meristems in grass and nettle

What obvious difference can you now see between the two plants?

All the meristems in grass are within a few millimetres of the soil surface, whereas in the nettle they are well above the soil.

The importance of this difference in relation to grazing is obvious. If an animal were to graze off all of the two plants above the dotted lines in Figure 23, this would leave no meristem in the nettle, but would not affect those of the grass. Remember that it is only from these meristems that the plant can *grow*. So now our grazed nettle cannot grow again, but the grass can regenerate quite happily. Very few large animals will graze closely enough to the soil to remove a grass meristem.

regeneration

There are plants other than grasses which have very short stems, for example, the daisies and plantains which are all too common on my lawn. However, most of these also have horizontal leaves, so that none of their structure is far enough off the ground to be grazed. They are thus *resistant* to grazing, unlike the grasses with their erect, easily grazed leaves. Plants like grasses are termed *grazing tolerant*.

grazing resistant

grazing tolerant

Because of the special structure of the grasses, a *sward* of grass (this is the term for vegetation composed largely or entirely of grass) can be cut or grazed several times during the year, and it will recover each time to be cut or grazed again. The grass sward receives continuous *inputs* of solar energy, carbon dioxide, etc., and the sheep take from it a more or less continuous *output* of leaf material. Other inputs to grass, as well as light and carbon dioxide,

sward

include: water from the soil, nitrogen from fertilizers via the soil, and a number of other soil nutrients which may have to be supplemented by the farmer. All these affect the *rate* at which the grass grows, so that, in general, the more nitrogen or other nutrients the farmer supplies, the faster the grass grows.

One of the important parameters of the grass compartment in our model of this system is the shape of the curve relating the rate of grass growth to the rate of supply of nitrogen or other nutrients. The shape of such *response curves* has been determined in a large number of experiments on grass grown in small plots. These are a good example of the general principle stated in section 4, that the parameters of any subsystem should, where possible, be determined in experiments conducted outside the system. The results of such experiments are illustrated in Figure 24.

response curve

The response curve is a very good example of a type of response known as *diminishing returns*, which is found for a large number of different systems, including financial as well as biological ones. The name 'diminishing returns' stems from the fact that the *return*, from a treatment, say fertilizer, does not increase by equal amounts for each succeeding increase in input rate (here, the increase in growth rate of grass for an increase in rate of fertilizer supply).

diminishing returns

SAQ 33

SAQ 33

Compare the increase in output rate as the input rate rises from 0 to 10 g of nitrogen per metre squared per year with the increase in output rate as the input rate rises from 30 to 40 g of nitrogen per metre squared per year.

The reason for this diminishing response is believed to be connected with other nutritional factors. Thus, although the grass may need more nitrogen to grow faster, it will also need more of other nutrients. If it is only supplied with extra nitrogen, there comes a point where its need for other nutrients is too great to be supplied in the existing situation. Giving it more nitrogen then has no effect, because its rate of growth is limited by the input rate of some other factor. Similar situations occur in nearly all systems with several inputs, as mentioned above. It you are interested in this topic, you can find more details of the fertilizer requirements of grasses in chapter 5 of Spedding's book *Grassland Ecology* and in Langer's book *How Grasses Grow.**

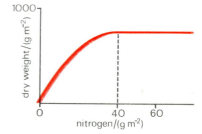

Figure 24   Response curve of plant growth with respect to application of nitrogen fertilizer

If a farmer has supplied his grass with all the fertilizer possible, its rate of growth will then be limited by the amount of light it can trap. As we saw in section 3.4, for individual separate plants, this depends on the size of each plant. When plants are growing close together, the situation becomes a little more complex, since each plant will shade others.

We shall now try to construct a simple model of how such a grass sward might grow.

SAQ 34

SAQ 34

Which of the techniques described earlier do you think would be most suitable for such a model?

5.6.2   *A simple model of grass growth*

This section takes you through the process of building a model in the form of differential equations. If you find this difficult, spend some time working through it, making sure you understand each step.

We shall construct our model using information which again has been obtained from experiments conducted on grasses either in the laboratory or under other controlled conditions outside the system. From such

---

* *Langer, R. H. (1972) How Grasses Grow, Edward Arnold.*

experiments, we know that the rate of growth of a leaf depends on the amount of light it receives. Every leaf, as does every living organism, needs some energy just to remain alive, without growing at all.

Exercise

Can you remember why this is so?

See section 1.2.

Thus, a leaf does not grow unless it is receiving at least a certain basic input of light. At this point, called the *compensation point*, the rate of energy input and the need for energy to maintain the leaf, just balance. Below this rate of input, the leaf will be using more energy than it receives, it will have to respire some of its own sugar and it will lose weight.

compensation point

When an area is covered with grass, sunlight falls on to the uppermost leaves, which absorb some of it, some passes through each leaf and some bypasses these uppermost leaves and passes through to the leaves below. Thus, less light reaches the lower leaves than the uppermost ones. If there are many layers of leaves the rate at which light reaches the lowest ones may be less than the compensation rate mentioned above, so that these leaves may be losing weight, while the ones at the surface are growing rapidly. Leaves grow at a rate proportional to the amount of light they receive. This, as we saw earlier, will depend on their size, and the amount of light available.

If we take w as the amount of leaf present at any one time, then the rate of growth can be written in calculus notation as

$$\frac{dw}{dt}.$$

If the average amount of light received by a unit of leaf is represented by l, we can state this formally in terms of a differential equation.

SAQ 35

Can you write out this differential equation?

SAQ 35

The amount of light received by a leaf will depend on the amount which has already been trapped from the sunlight by the leaves above it. Thus for the leaves nearest the top the rate of light input is equal to the rate at which it arrives from the sun. For those at the bottom it will be this rate, minus the rate of trapping by all the higher leaves. This rate of trapping will be equal to some constant times the amount of leaf in the whole canopy, or, using symbols, $s \times w$, where s is a constant representing the rate of trapping per unit of leaf. We can now write an expression for the light l received by the average leaf, in the middle of the canopy.

Can you now write this expression?

$l = $ (in-coming light above canopy) $-$ (in-coming light) $\times s \times \frac{1}{2}w$.

Remember that we are assuming half the total leaf is above the *average* leaf.

If we represent the in-coming light by I, we can now put this into the form

$$l = I - \tfrac{1}{2}Isw$$

$$= I(1 - cw), \qquad \text{where } c = \tfrac{1}{2}s.$$

Now we can combine this with the earlier expression for $\frac{dw}{dt}$ and, assuming that I is constant for any time, we get the formula

$$\frac{dw}{dt} = k \times w \times I(1 - cw).$$

Since I and k are both constant we can combine them into one constant K,

$$\frac{dw}{dt} = Kw(1 - cw). \tag{6}$$

Now, if the amount of leaf present increases, ultimately the lowest leaves will be practically in the dark and will be below their compensation point. The upper leaves will still be growing but the lowest leaves will be losing weight. In the end a point will be reached where there are so many leaves losing weight that their weight loss will equal the weight increase of the illuminated leaves. The total weight of the canopy will then remain constant.

SAQ 36

What does this mean in terms of the symbols above?

If we represent the maximum size that the canopy ever achieves by m, we find that the constant c from equation (6) above is equal to $1/m$. This gives us the conventional form of the *logistic equation*. This is

$$\frac{dw}{dt} = Kw\left[1 - \frac{w}{m}\right]$$

or $\quad\dfrac{dw}{dt} = Kw\dfrac{m - w}{m}. \tag{7}$

This equation forms a very useful first approximation to the growth of any organism in an environment where one factor, such as light, is in limited supply. Of course, it *is* only an approximation for the growth of a grass crop but, accepting this limitation, we can use this model to look at a simple problem in grazing management.

In addition to the input of grass by growth there will be an output of grass to the grazing sheep. This will clearly affect the amount of grass present, since we could regard the grass compartment as being the same as the bucket in the example in section 3.2.1.

Exercise

Draw a compartment model of the grass.

See Figure 14. The only difference is that we do not know the values of the input and output rates.

In a very simple grazing situation, we might want to ensure that we obtained the maximum sustainable rate of output from grass to sheep. Can we devise a scheme for doing this, if we assume that the input of growing grass is represented by equation (7)? Clearly, if we require the maximum *sustained* output of grass, we must also maintain a constant amount of grass present, since, if the amount present falls, we shall ultimately run out of grass, and, if it rises, we are clearly not taking the maximum possible output.

Exercise

What does this imply, in the terms of equation (7)?

That w is being maintained constant. To keep the amount present constant, the rates of input and output in the compartment must be equal. So, to obtain the maximum output, we must choose the value of w which will give maximum input.

Knowing the parameters of equation (7) we can use a simple mathematical technique to find the value of w for which $\dfrac{dw}{dt}$ is greatest. In this equation, the rate of growth is expressed as a *function* of size. In general, if we have any function — $f(w)$ in this case — we can find the value of the independent variable for which the function has its *maximum value* by differentiating the function with respect to the independent variable and equating this to zero. Here we want to find $f(w)$ and to solve the equation $f'(w) = 0$.

Differentiating equation (7), what do we obtain? (Refer to the table in the Mathematics Booklet if in doubt.)

$$f(w) = Kw\frac{m-w}{m}$$

$$= Kw - \frac{Kw^2}{m},$$

$$f'(w) = K - \frac{2Kw}{m}.$$

The maximum will be when $f'(w) = 0$, i.e.

$$K - \frac{2Kw}{m} = 0,$$

$$\frac{2Kw}{m} = K,$$

$$2w = m \quad \text{or} \quad w = \tfrac{1}{2}m.$$

The maximum input rate is obtained if we keep the amount of grass at half the maximum possible value. This model and its solution describe the simplest possible situation. In practice we shall need a more sophisticated model of our grass canopy to allow for changes in the rate of growth with the age of leaves, to allow for changes in temperature, sunshine hours, etc. However, you can see that the *principles* on which such a model would be constructed do not differ from those used here. The actual equations would be more complex, to take account of the larger number of input variables, and their solution would almost certainly involve using an analogue computer or some approximate method. This illustrates how the power of differential-equation models, with their associated theory, can be used in a practical problem.

5.6.3 *Other grazing plants*

Our model here assumed that the only plant involved was grass, and that conditions were uniform at all times. In practice this would not be so, since we almost always find some plants other than grasses in grazed fields. One particularly important category are the *legumes*. These are a different family **legumes** from the grasses, and include grazing plants such as the clovers, lucerne and sainfoin as well as more familiar garden ones like peas, beans and laburnum. Their importance in grassland stems from the fact that they are capable of using gaseous nitrogen from the air as a source of nutrient, making them independent of applied nitrogen fertilizer for their growth. They accomplish this with the aid of bacteria living inside the roots of the legumes, which trap the nitrogen and convert it to a form that the legume can use, receiving in exchange sugars from the leaves of the legume.

SAQ 37

SAQ 37

What type of nutrient-intake process is this?

When the roots of legumes die, they are broken down by saprophytes in the soil and their nitrogen is released into the soil, where it can be taken up by the roots of grasses. In addition, the leaves of legumes usually have a very high protein content, which is ideal for the animals grazing on them.

A typical plant of white clover is shown in Figure 25. Like most plants found in heavily grazed areas, it keeps its meristems near the soil surface, where they are unlikely to be removed by grazing, although the leaves are borne on long stalks (called petioles) and are accessible to the grazing animals.

The transfer of nitrogen from the air, through legumes, to grass can provide the equivalent of about 100 kg of nitrogen per hectare every year.

Figure 25   A typical plant of white clover (meristems marked in red)

Most farms only apply about this much nitrogen as fertilizer, so that the importance of legumes in the nitrogen flow processes of the ecosystem is clear. The grass–legume subsystem is one which has been modelled using differential equations such as we have discussed earlier, details of the model are given in reference 4 of section 3.6.

In our simple model, we also assumed that solar energy input was constant, and that the grass plants did not change with age. In practice we receive more solar energy per day in summer and the temperature is also higher, so that grasses grow faster in summer than in winter. In addition, they grow much faster at flowering time, around May and June, than at other times of the year and at different rates in different areas. The daily growth rates of two typical species of grass on three sites are shown in Figures 26 and 27.

From these you can see the rapid growth in May and June, and the much slower growth throughout the year on the hill sites than on the lowland site. You will notice that the growth rate in winter is plotted as zero. In fact grass does grow even then in the lowlands, but too slowly to be measured in the

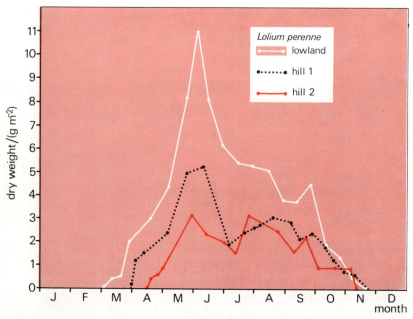

Figure 26   Daily growth rates of Lolium perenne (*perennial ryegrass*) *on three sites*

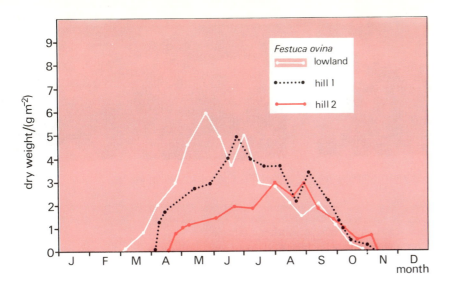

Figure 27   Daily growth rates of Festuca ovina (*sheep's fescue*) *on three sites*

experiments from which these curves were drawn. Thus the output from grass to the sheep must vary considerably during the year. The effect of this variation on sheep management will be discussed later, and is illustrated in the second television programme associated with this module.

Next we shall look at the parameters and variables associated with the second compartment, the sheep.

5.6.4   Sheep

A selection of the range of types of sheep found in the United Kingdom is shown in Figures 28–32 ranging from the primitive Soay, which runs wild in some of the Scottish Islands, to the Colbred, a breed recently developed for special purposes.

The essential feature common to all sheep is that they are *ruminants*. This means that instead of a simple stomach, like that of man and other mono-gastric animals, they have a complex collection of 'stomachs', as shown

Figure 28   The primitive Soay sheep, which runs wild on some of the Scottish Islands

ruminant

Figure 29   Scottish Blackface, found mostly in the Scottish uplands

Figure 30   Dorset Horn, a breed of the southern counties of England

Figure 31 *Merino, a comparative rarity in the United Kingdom, but the major breed in Australia*

Figure 32 *Colbred, a specially developed hybrid*

in Figure 33. In the largest of these sacs, the *rumen*, we have another example of symbiosis. The fluid in the rumen contains a mass of bacteria and protozoa (other micro-organisms). These can break down the cellulose cell walls of plant material eaten by the sheep to form *fatty acids*, taking for their own use some of the energy liberated and releasing the fatty acids to be absorbed and utilized by the sheep as an energy source. If we tried to eat grass ourselves, being monogastric animals, it would pass directly through our gut undigested, apart from the small amount of sugar which we could absorb.

rumen

fatty acid

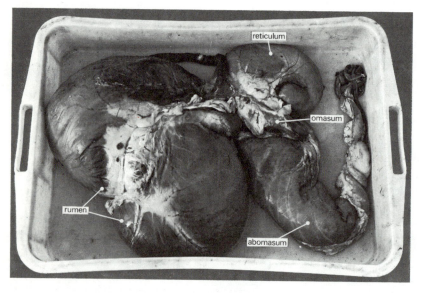

Figure 33 *The 'stomachs' of a sheep*

Sheep's jaws are also specially adapted for grazing. In place of the front upper teeth (incisors), they have a *dental pad*, against which the lower teeth chop grass, rather like some secateurs. The jaws are shown in Figure 34.

The products useful to man, from sheep, are the meat, wool and, to a lesser extent, the skin. Butchers cut up the carcasses according to the scheme shown in Figure 35. As you know, the most expensive joints are from the hindquarters, followed by the shoulder, so the farmer, and butcher, both look for an animal with the highest possible proportion of meat in these areas relative to the less desirable parts like neck and ribs.

Wool is used mainly for clothing or carpets, depending on its fineness, degree of waviness, length, etc. Some examples of different wools are shown in Figure 36. Some sheep have a proportion of hollow *hairs* in their fleece as well as solid wool fibres. This is usually undesirable, and such mixed fleeces fetch a lower price.

The input to the sheep compartment is mainly grass. Its output, as we have seen, is meat and wool, plus dung. Sheep reproduce usually only once each year. The lambs are sold for slaughter only after they have grown to a certain size, and so the output of meat from the system is seasonal. Similarly, shearing occurs only once annually. This seasonal nature of the output from the system causes some problems in economic terms, as we shall see.

Most sheep will mate only in autumn. Their reproductive cycle is controlled

Figure 34   The jaws of a sheep, showing the dental pad on the upper jaw

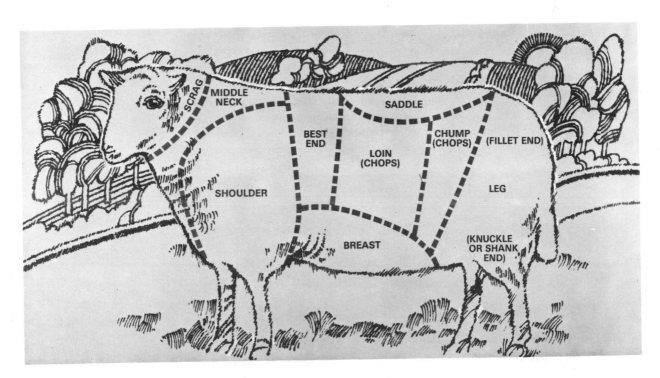

Figure 35   The 'cuts' of lamb

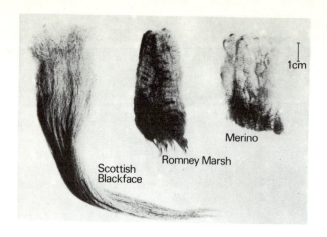

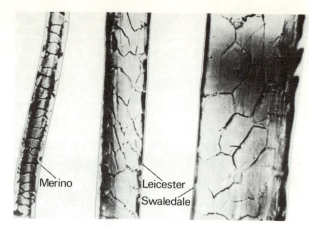

Figure 36(a)   Samples of three types of wool

Figure 36(b)   Highly magnified views of individual wool fibres of three types

by the changes in day length and they are usually only fertile when mean day length is less than twelve hours and decreasing. Lambs are born after a gestation period of 147 days. Most common breeds of sheep only produce one, two or three lambs each year, although there are one or two special breeds which have up to seven lambs at a time.

After their birth, the lambs suckle their mothers for a variable length of time during the summer, at the same time eating gradually increasing amounts of grass. When they reach a weight of 25–30 kg, the majority are sold for slaughter. If, for some reason, they have not reached this weight by the autumn, they are usually sold as *stores* to other farmers who fatten them on other crops or cereal rations to slaughter weight. In addition, some ewe lambs may be sold to other farmers as breeding stock.

stores

Clearly, the ewes which bear the lambs do not live indefinitely, and their fecundity declines with age, so that after a life of from five to eight years they also are usually sold. To replace them the farmer either has to buy ewes from other farmers or keep some of the ewes' female offspring each year to maintain the breeding flock.

SAQ 38

SAQ 38

What modelling technique could we use to simulate this situation.

We shall try to set up a model of a sheep flock, using a transition matrix. If you are in any doubt about these, refer back to section 3.5. You can set this up on your computer terminal and, by varying the model parameters, you can try to find what proportion of the lambs born each year should be kept to maintain a stable flock size. A similar model could also to used to determine to what age the ewes could be kept before the overall fecundity of the flock became too low. To determine this would require extra information on the cost of keeping a ewe for a year, the price per lamb and the price for which an old ewe could be sold.

What information is required to set up the simple matrix model?

The information required is the probability of survival of female sheep from year to year and the number of female young born to sheep in each age group.

Suitable information to construct such a model, assuming that the ewes are kept until they are five years old, is given in Tables 3 and 4.

Table 3   Average number of female lambs born per ewe in each age group

Age/years	0–1	1–2	2–3	3–4	4–5
Females born	0	0.9	1.2	1.0	0.9

Table 4   Average survival from one age class to the next

From age	To age	Probability of survival
1–2	2–3	0.98
2–3	3–4	0.95
3–4	4–5	0.80
4–5		0.00

The above figures are fairly typical values for a flock in Britain. You will note that no probability of survival has been given from age class 0–1 to age class 1–2. This you must supply yourselves, remembering that the percentage of lambs sold will represent $100 \times (1 - \text{probability of survival})$.

SAQ 39

SAQ 39

If we sell 90 per cent of the lambs and there is no other source of loss, what is the probability of survival from age 0–1 to age 1–2?

Assuming we have an initial flock of forty ewes, with ten in every age class, and forty lambs (0–1 year old), the population vector will be

$$\begin{bmatrix} 40 \\ 10 \\ 10 \\ 10 \\ 10 \end{bmatrix}.$$

Assuming the sale of 95 per cent of the female lambs, the appropriate transition matrix will be

$$\begin{bmatrix} 0 & 0.9 & 1.2 & 1.0 & 0.9 \\ 0.05 & 0 & 0 & 0 & 0 \\ 0 & 0.98 & 0 & 0 & 0 \\ 0 & 0 & 0.95 & 0 & 0 \\ 0 & 0 & 0 & 0.80 & 0 \end{bmatrix}.$$

Carrying out the appropriate multiplications,

$$\begin{bmatrix} 0 \times 40 + 0.9 \times 10 + 1.2 \times 10 + 1.0 \times 10 + 0.9 \times 10 \\ 0.05 \times 40 + 0 \times 10 + 0 \times 10 + 0 \times 10 + 0 \times 10 \\ 0 \times 40 + 0.98 \times 10 + 0 \times 10 + 0 \times 10 + 0 \times 10 \\ 0 \times 40 + 0 \times 10 + 0.95 \times 10 + 0 \times 10 + 0 \times 10 \\ 0 \times 40 + 0 \times 10 + 0 \times 10 + 0.80 \times 10 + 0 \times 10 \end{bmatrix} = \begin{bmatrix} 40 \\ 2 \\ 9.8 \\ 9.5 \\ 8.0 \end{bmatrix}.$$

Repeating this several times, we find that, even allowing for the problem of interpreting what we mean by 9.8 sheep, the total number of ewes is declining. You can find the correct selling percentage using the programme on your computer terminal, trying different values for first-year survival.

5.6.5   Grazing management

Lambs are only present as graziers during the summer. Thus, the input of grass which is required by the sheep compartment in our model will vary through the year. Measuring the daily amount of food required by some typical flocks of ewes, with different numbers of ewes on a given area of ground and with different numbers of lambs, has given curves like those shown in Figure 37.

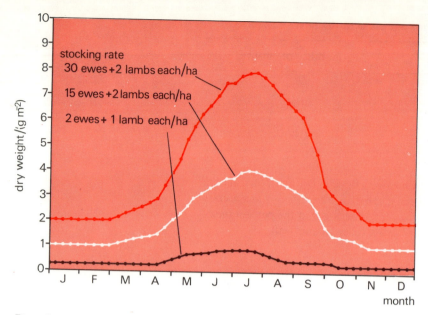

Figure 37 Daily requirements of sheep for grass at three stocking rates

Exercise

Compare these curves with Figures 26 and 27 for the daily output of grass. What differences can you see?

For clarity, the two sets of curves, the output of grass and the required input for the sheep are shown on the same axes in Figure 38.

In winter, the output of grass is less than the input required by the sheep. In spring, the converse is generally true.

How can these differences be reconciled? One way would be for the farmer to keep only as many sheep as could obtain their required input during the winter from current production, plus any surplus production from the summer which was still present. Alternatively, he might be able to buy enough feedingstuffs to feed his sheep during the winter and let them graze all the grass available during the summer. A more common approach, however, is to make use of the food store shown in Figure 18. In spring and early summer, when the output from the grass is greater than the sheep's input need, the surplus is directed into the foodstore, where it is kept until winter, when the sheep's needs exceed the amount of grass available. Then they are fed the stored food, until the following year's grazing becomes sufficient for their needs.

Grass is stored in one of two ways. These are as hay and as silage.

Hay is grass which has been cut and allowed to lose water until it is dry enough to store without rotting. Silage is cut grass which is allowed to ferment slightly, in the same way as yoghurt, after which it will remain without further decomposition for a long period of time. Use of stored food, plus some purchased feedingstuffs, enables the farmer to match fairly closely the nutritional requirements of his sheep with the grass growth which is available. Any discrepancies which remain are smoothed out by the animals' reserves of body fat, which are built up during periods of good feeding and can be used during periods of scarcity. This is discussed in more detail in chapters 8 and 13 of Spedding's *Grassland Ecology.*

In addition to the overall problem of matching the general needs of the sheep to the supply of grass available, the farmer also has to consider some specific problems. Grass, if left to grow without grazing for a long period, becomes tough and difficult for a sheep to digest. Also, if it is grazed too often, the combination of small plants trapping only a little sunlight and frequent

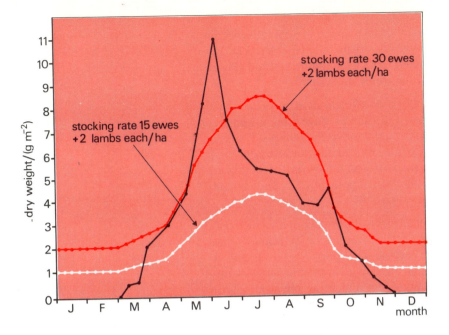

Figure 38(a)   Daily grass requirements of sheep compared with daily growth rates of grass on a lowland site

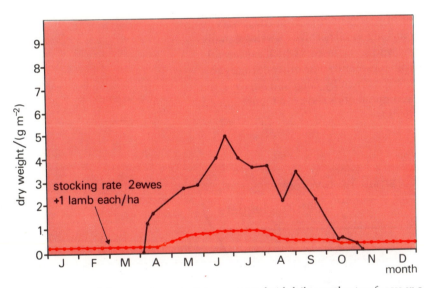

Figure 38(b)   Daily grass requirements of sheep compared with daily growth rates of grass on a hill site

trampling by the feet of many sheep can slow the growth rate of the pasture markedly.

The ideal grazing management in practical terms involves separating the available land into small units, each of which in turn is grazed by the flock and then allowed to rest and recover for a period. Such units may be any size from one to ten hectares (1 hectare = 10^4 m^2) and each of these is grazed for between two and ten days, then rested for five or six weeks. This not only benefits the grass, but helps to get rid of some of the sheep's intestinal parasites whose eggs are shed along with the sheep's faeces and can then re-infect any sheep which eats the eggs within a month or so. Spedding deals in some detail with these parasites in chapter 13 of *Grassland Ecology*. Against these advantages, division into small blocks may be very expensive on hill land. Here, much less control is possible over the grazing animals, so that in general they are much less productive than those on lowland pastures.

5.6.6 *Soil*

The last compartment in our general model of our ecosystem is the soil. This includes not only the non-living clays, silts, organic matter, water, etc., but also the small animals, fungi and bacteria which inhabit the soil and play an essential part in the functioning of the whole ecosystem.

The non-living parts of the soil serve two main functions in the system: they act as a support on which plants grow and animals move around, and they form a reservoir of nutrients which can be used by living organisms. The three most important nutrients are nitrogen, phosphorus and potassium, the N, P and K of garden fertilizers. Nitrogen is stored largely in the non-living organic matter of the soil, and phosphorus and potassium as solid minerals and solutions.

N P K

In addition to its role as a reservoir of nitrogen, the organic matter, or *humus*, in the soil helps to stabilize soil structure. In the absence of sufficient humus, soils tend to break down and become non-porous. This leads to waterlogging and a whole lot of problems in cultivation and with diseases. Grasses whose leaves are continually dying and being replaced help to maintain and increase the organic matter in the soil. Added to this, the animals grazing the grass defaecate upon it and the faeces are added to the organic matter in the soil, as well as providing a food supply for some of the soil organisms.

humus

Too much dead organic matter can be a disadvantage if there are not enough of these soil organisms to break the dead material down into simple components and release the nitrogen which can be utilized by growing plants. Where the soil organisms are insufficiently active, the dead leaves and other materials build up in the form of *peat*. The organisms necessary to break down the dead organic matter include small herbivores – slugs, snails, etc. – detritus feeders like earthworms, and some insects which eat recently dead organic material. The excreta from these, and some other dead organic material including these animals is broken down by *fungi* and *bacteria*, the decomposers. The soil system with its components is shown in Figure 39, and in chapters 9 and 12 of *Grassland Ecology*.

peat

decomposers

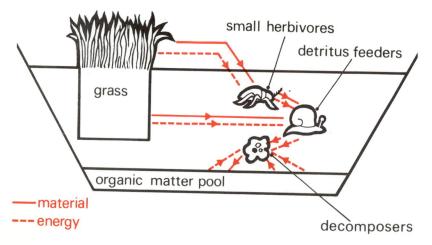

Figure 39   The soil system and its components

5.7   Overall systems diagram

We are now in a position to fill in all the major inter-relationships in the whole system, shown in schematic form in Figure 40, in terms of the energy and material flows between compartments. This gives a picture of sheep production as an ecosystem. However, we said before that it also had features of two other systems.

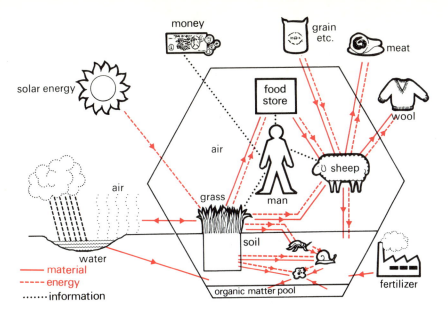

Figure 40   The whole system of sheep production

SAQ 40   **SAQ 40**

What were these two other systems?

We shall next look at some aspects of these two systems as they impinge on sheep production.

Exercises

Why is a food store necessary in the sheep production system?

See p. 58.

What is the role of the soil in the system?

See p. 60.

6 Interaction with other systems

6.1 Economic systems

One of the ways in which a sheep farmer can make good the shortfall in grass output during the winter period is to buy feedingstuffs from other farmers or from merchants. This is a money transaction, so the farmer obviously needs cash to pay for these feedingstuffs. As we noted earlier, the output from the sheep-production systems is highly seasonal, so that the farmer only has something to sell for cash in early summer (wool) and in late summer/autumn, when the lambs are fit for slaughter. His need for purchased feedingstuffs comes in winter, so he must therefore either retain enough cash from his sales to pay for the feedingstuffs or obtain credit against his future sales. This seasonal imbalance in cash flow is a feature of most farming systems, where production is tied very closely to the season of the year.

In addition to the seasonal nature of the output from the system of sheep production, there is a problem of risk.

An unusually cold winter, disease or simply an error in estimating grass growth can all lead to a severe cut in the output of meat and wool from the system. Because of the unpredictability of these abnormal inputs to the system, anyone lending money to be used in the system expects a higher interest rate to cover the risk involved. This is one area where a well-quantified model of the whole system of sheep production would be valuable, so that the overall response to abnormal inputs could be estimated more reliably. In the absence of such models, the risk is usually given the highest possible value in considering interest rates (just in case)!

Here we see a very good example of the interaction between the properties of the system in biological terms and its needs in economic terms.

6.2 Socio-technical systems

Social systems also influence the success or otherwise of sheep production, though this influence may not always be obvious. By definition, the system must be located in a rural area, with a population density lower than the national average. This means that social amenities such as schools, shops, bus services, etc. must each cater for fewer people at higher cost per individual, or must serve large areas, where the individual pays the cost of transport. Sheep farmers suffer particularly from these disadvantages, since many such farms are in very remote areas, where sheep are the only productive animals which can be used.

All societies moving towards industrialism have relied for their labour on a steady transfer of labour from farms to other industries. This transfer is accelerated when the social amenities of the rural areas not only fail to keep pace with those of urban areas, but actually decline both in quality and economy. The end result of such a process could be that only those with insufficient skill or education to obtain industrial employment are left behind to manage an extremely complex ecosystem.

Counter to this, there is clearly an element of job satisfaction, as defined in Module 3. A shepherd's job may be physically demanding, but is rarely repetitive – a complaint levelled against many urban jobs. In addition, many people find the absence of some urban amenities attractive in itself. Thus, there is clearly a demand for access to rural areas for recreation; the integration of recreational with productive use of land is currently a pressing problem in many areas.

One further social problem is worth mentioning. Many sheep farms are what is known as *family farms*. Here most or all the labour is provided by members of the family, without the employment of outsiders. Over the whole sheep industry, the input of labour available is relatively constant, or possibly declining steadily. However, within each family, the labour available will vary much more drastically. A young newly married couple may both undertake a considerable amount of outdoor work. However, when children arrive, the wife's farming activities will be curtailed, so the labour input may be reduced by up to one third, and the demand for cash for 'non-productive' activities will increase. Again, when the children reach their mid-teens, there may be a surplus of labour available with both parents and children active. At this stage, many children take jobs off the farm, which prove more attractive, and ageing parents are then left to cope on their own. This alternation between over- and under-provision of labour may be counterbalanced where hired labour is available. In remoter areas, hired labour may not exist, and the fortunes of each farm tend to go through a cycle depending on the family labour available.

family farms

SAQ 41

SAQ 41

Can you identify the important inputs to the system of sheep production?

7 History and possible future development of the system

Here we deal with the last of the systems questions given in section 2.3.1. You will have an opportunity of seeing both some of the history and some new developments of sheep production in the television component of this module. A brief outline of the important stages in the history of sheep production is given below.

The development of present systems of sheep production has occurred over a very long period of time, the most rapid change taking place during the eighteenth and nineteenth centuries, with odd developments prompted by war-induced shortages of imported meat at intervals during the present century.

The two most important steps in development of present systems involved firstly the enclosure of common grazing land, and secondly the genetic improvement of sheep and grass. With common ownership of grazing, control over mating, grazing method and other variables, was virtually impossible. With the enclosure of fields as we know them, during the eighteenth century, some degree of control became possible, and *selective breeding* (which, interestingly enough, was first practised among race-horses and fighting bulls, rather than productive stock) could be applied to sheep. Examples of successful exponents of these techniques were Robert Bakewell, Thomas Coke and other eighteenth-century landowners. Their efforts were important in modifying the *selective pressures* in unenclosed areas, which otherwise tended to favour the most energetic and most aggressive animals, rather than those which produced the most suitable carcasses. Similarly, during the late nineteenth century, when a market in grass seeds developed, the seed available tended to come from grass selected *by* its seed bearing capacity, rather than *for* its output of useful leafy material. This highlights again the distinction between managed and unmanaged ecosystems, the sole 'purpose' of the latter being the attainment of the individual goals of survival and reproduction of species.

selective breeding

selective pressures

Techniques for storing grass developed most rapidly after some progress had been made in selecting grasses which produced a useful surplus to store. Grazing methods, likewise, have developed gradually over the years since then, although there may be a considerable lag between the suggestion of a new technique by research workers and its adoption by the majority of the industry. This is due in part to a natural conservatism in the operators, but is also due to the long lag periods involved in the system, between receipt of an input, such as a new technique, and seeing a measurable change in output.

lag

Currently, a number of new techniques are being investigated, at a research level. Among the more promising are methods of controlling the breeding performance of the ewe, making the conception of a defined number of lambs at a specified time more certain. Presently, the limitation in litter size to a maximum of about four is a major bottleneck to the production of meat and wool. Similarly, the very pronounced day-length requirements for mating in sheep presents the problem of seasonal shortage/surplus of lamb on the market.

controlled breeding

Improvements in the genetic make-up of the sheep continue, as does the introduction of new drugs for the control of parasites and disease.

The importance of the legumes in grazing management was illustrated in section 5.6.3. The problems of water pollution due to rainfall after heavy application of nitrogen fertilizer and the cost of such fertilizers under EEC conditions has given fresh impetus to research into better legume varieties, looking for legumes which trap larger quantities of nitrogen, are compatible with grasses in a mixed sward and do not contain any toxic or deleterious compounds. It seems more sensible to obtain a limited, but more or less free supply of nitrogen from legumes, rather than to pay for a potential pollutant.

Storage of grass has always been a problem, either as hay or silage. One interesting development that has received some attention is the possibility of using a press to squeeze the juice out of grass, rather than allowing grass to dry naturally or in a forced-air drier. The residue left after pressing can be completely dried very easily, and has about 80 per cent of the feeding value of the original grass. In addition, the juice extracted has a very high protein content and after curdling, can be made into a 'cheese' which is ideal for feeding to pigs, or even to humans. Use of this technique could increase meat production from grass very considerably.

However, perhaps the area in which most development is likely to occur over the next few years is the *uplands*, where improvements in grazing management and possibly the use of new methods for feeding sheep in winter could increase dramatically the production of meat. They are presently little advanced from an unmanaged condition, due to the problems of farm size, common ownership of grazing, and an uncertainty among both farmers and land use planners as to the future use of such areas. The conflicting claims of sheep, forestry and recreation create a very difficult problem to solve. Piecemeal solutions have been applied, as a result of considerable pressure from individual interests, but an overall realistic model even of individual enterprises has yet to be constructed. It is in this area that systems techniques are most needed, to obtain a rational solution of the problem. Hopefully, courses like *Systems Behaviour* will be able to help in this.

upland areas

8 The role of systems techniques in sheep production

We have seen an outline in this case study of how meat and wool are produced. In the television programme associated with this module, you will have an opportunity of seeing some practical examples of systems of sheep production. Many of the details of day-to-day management of sheep may seem rather confusing. It is helpful to realize that these systems are all examples of our basic compartment model, that shown in Figure 40. At the present stage of development of models of ecosystems, we cannot present a fully quantified version of this, with all the variables and parameters specified. However, a simplified picture of the energy flow system is given in the Broadcast Notes, with values of the major variables for the systems shown on the television. This is the first use which we can make of a systems approach to sheep production, as an aid to our understanding of the apparent complexity.

A second use for systems models in sheep production, and in dealing with ecosystems in general, is as a substitute for large-scale experiments on whole systems.

SAQ 42

SAQ 42

Why are these large-scale experiments undesirable?

For the individual farmer, a fully quantified model of his sheep enterprise could help as a management tool, to decide, for example, how much fertilizer it would be worth while to apply, or whether or not to adopt some new technique. Here again, we could determine the answer to these questions by experiment, but the problems of cost etc. are tremendous. Clearly, many farmers do make simple models of their enterprises, either mentally, or on 'backs of envelopes'. The successful ones may or may not be the ones who make the clearest models!

At the present time, however, the greatest use of systems models is in agricultural research. It soon became clear that experimentation with complete farm systems was largely impractical, in the same way as were experiments on any whole ecosystem. As a result, most research, both in agriculture and ecology generally, became more analytical, studying smaller and smaller subsystems in greater detail. If such experiments were regarded as providing data on the parameters of a subsystem in an overall model, this would be extremely useful. All too often, the experiments were conducted as a result of the curiosity of an individual, or of the availablity of material or an experimental technique. This became clear when the first systems scientists actually tried to assemble quantitative models of whole systems; some subsystems were extensively documented, others had never been examined at all.

Ideally, experimental research and theoretical modelling should complement one another; this is certainly the view of the author of the Appendix to this module.

A fuller discussion of the role of modelling in ecological research generally can be heard in the radio programme for this module. You should try to form your own conclusions after listening to this and reading this module.

Glossary

Sheep farming has its own language or languages, spoken in different parts of the United Kingdom. A glossary of some of the terms used is given below.

Ewe	– Female sheep which has borne at least one lamb
Hogg, hogget, teg, gimmer, theave	– Female sheep prior to first lambing
Tup, ram	– Uncastrated male sheep
Wether	– Castrated male sheep
Flushing	– Extra feeding given prior to mating to increase conception rate
Crutching, tailing, dagging	– Cleaning off soiled wool from hindquarters
Dipping	– Control of external parasites by immersion of sheep in insecticidal bath
Fluke, footrot, nematodirus, roundworms, strike, flystrike, mastitis	– Sheep disease
Gathering	– Collecting sheep together off open mountain grassland
Hefting	– Hereditary attachment of sheep to particular area of mountain, so they do not stray voluntarily

Appendix

Mathematical models in ecology: an agricultural research scientist's point of view

N. R. Brockington, Grassland Research Institute, Hurley

[From Jeffers, J. N. R. (ed.) (1972) *Mathematical Models in Ecology. 12th Symposium of the British Ecological Society*, pp. 361–5, Blackwell.]

A farmer is essentially a manager of whole biological systems: he has a holistic view of biology from necessity. The agricultural ecologist adopts a similar viewpoint because he is interested in the inter-relationships between plants and animals and their environment; he appreciates that any one component of a system cannot be described properly in isolation. This attitude brings both the farmer and the ecologist face to face with the same problem, the complexity of biological systems. Analytical research confronts them with more and more information on the individual components of these systems and the urgent need in applied research is to make sense and use of all these data.

This excellent and wide-ranging symposium has served to strengthen my personal belief that a range of powerful and appropriate tools is available to us for this task in the form of mathematical models. It has been well demonstrated, if that was necessary, that mathematics has the important advantages of brevity, precison and ease of manipulation, for the job of describing and studying complex ecological and agricultural systems. In his opening address, Mr Jeffers stated that the challenge to ecologists is whether they will be willing to take advantage of these tools. I wish to return to his interesting point that this challenge should be particularly related to research strategies, rather than only the tactics of research. Meanwhile, accepting the general proposition that mathematical models can provide the tools we need, there remains a number of important questions around which many of the papers and discussions have centred. Which tools are most appropriate for particular jobs? How does one use them? What, precisely, can they do and what can they not do? Finally, what are the specific risks and dangers inherent in their use?

The choice of tools

One may be sure that where two or more model-builders are gathered together a spirited discussion of alternative techniques will occur! The symposium has confirmed this general rule, but in a new and rapidly developing field of work it is both necessary and valuable to examine the tools of the trade in some detail, and I believe we have had our thoughts clarified and our horizons widened in this respect. The attention given to 'event-orientated' formulations which do not appear to have received the consideration they deserve in biology, has been especially valuable. Similarly the use of models incorporating stochastic variation has been rightly emphasized.

The importance of models being understandable to field biologists has been emphasized by a number of speakers, and this is particularly relevant when considering which tools to use. I hold unashamedly to the view that simplicity of operation is one of the most important criteria. The biologist is employed to solve biological problems, and the simpler the tools he has to use and to explain to his colleagues, the better. If he is obliged to spend much of his time on the mechanics of using a tool, he will be in danger of losing sight of the original problem. If he asks an 'expert' to do the whole job for him, this can lead to difficulties of communication, misunderstandings and errors. The examples which have been described of successful team projects, involving both biologists and mathematicians, are very encouraging; but simplicity of formulation must be equally, if not more, important in these circumstances.

The problem-orientated simulation languages of which we have heard are important if one is seeking for simplicity. The suggestion that the special attributes of analogue computers may be harnessed via digital machines, without being directly involved in the complexities of the electrical circuits, is also relevant.

Using the tools

The fact that models are, by definition, simplifications of reality has been a recurrent theme of the symposium. Nevertheless, it is worth reiterating that if a model were to be an accurate representation of reality, complete in every detail, then clearly it would be essentially equivalent to the real object. Because it faithfully mimicked every detail it would be no less complicated, no less difficult to understand or to manage. This truism illustrates the principle that simplification is not only usual but essential.

Apart from the practical considerations of whether it is feasible, with the technical resources that are available, to build complete models of large and complicated biological systems, there appears to be some danger of creating 'monsters'. Even the creator of one of these may not fully comprehend it, and could end up simply pushing buttons and producing results which cannot be used sensibly.

There is an equal danger, at the other extreme, of over-simplification, leading to models which are easy to understand but which are valueless or even positively misleading.

To arrive at a satisfactory compromise, the importance of a clear definition of the purpose of a model has been noted by a number of speakers. A vague definition of the objective on the lines of wanting to understand the system is not sufficient, and can only result in trying to build a model which is all things to all men. Even if it is successful in the technical sense, such a large and complicated formulation will be so cumbersome as to be virtually unusable. A limited, and clearly defined objective is essential in order to judge how much simplification can be tolerated for that particular purpose.

An integral part of the process of deciding how much detail can be sacrificed in the interests of clarity and usability is to check whether the behaviour of the model is in sufficiently close agreement with that of the real system; in other words to test or validate the model. Clearly this cannot be attempted without a closely defined objective for the modelling exercise.

A major problem in model testing which has emerged in a number of discussions is the danger of 'circularity'. In the simplest case, the trap for the unwary model builder is that he will be tempted just to confirm his arithmetic, or the computer's arithmetic, by testing the model with the identical data which have been used in its construction. De Wit's rigorous formula for avoiding this problem* involves constructing the model from data at a different, more detailed, level of understanding from that at which it is tested e.g. the use of data from physiological experiments under controlled conditions to model crop growth in the field. The corollary of this is that, when there is a discrepancy between predicted and measured behaviour of the system, one resorts to further experimentation at the detailed level of understanding to improve the model. In this way, the independence of the data used in construction and testing can be guaranteed. However, this is perhaps

* de Wit, C. T. (*1970*) '*Dynamic concepts in biology*' *in* Jones, J. G. W. (*ed.*) The Use of Models in Agricultural and Biological Research, *pp. 9–12, Grassland Research Institute.*

an ideal solution, which cannot be achieved invariably in practice. It has been suggested that an alternative solution may be to draw the data for construction and testing from the same level of understanding, perhaps from the same field situation, but on different occasions in time e.g. once data has been accumulated for, say, one growing season it may be used legitimately to predict model behaviour in subsequent seasons. While there may be some doubt as to the philosophical validity of this procedure, perhaps a more important question is whether it is an efficient use of available resources in practical, day-to-day model building.

Attempting to answer the more specific question, how to use models in applied agricultural research, the choice of an appropriate degree of detail in the formulation is especially relevant. Much agricultural research has relied on simple input/output experiments in the field. The main justifications for this approach is that it ensures, as far as possible, that the results can be directly applied in farming practice, since the trials are carried out under field conditions which closely resemble those on commercial farms.

Alternatively, more detailed experiments can be carried out, in the laboratory, greenhouse or animal house, on specific components and relationships within an agricultural system. In this case, one has much more control of the experimental material and the treatments imposed on it, and the hope is that such work may help in elucidating the mechanisms which contribute to the overall behaviour of the system.

Both these approaches appear to have inherent limitations in applied research. In field experiments, it is seldom possible to do more than measure the treatment inputs and the gross outputs in terms like final crop yield or animal weight changes. The results can only indicate what happened in one set of ill-defined circumstances and not why or how: one has a 'black-box' situation. If the farmer is to manage the system efficiently, he needs to know the effects of the many possible treatment variations and combinations, under varying conditions of climate and weather, soil type, etc. But, with the sort of information which the field experiment produces, it is dangerous to attempt to extrapolate, or even interpolate.

Consequently, a vast number of field experiments are necessary to cover all the possible treatment and site variations. This is not to belittle the very considerable progress which has been made in scientific agriculture using this technique, simply to recognize its inherent limitations.

The use of more detailed experiments under controlled conditions is equally open to criticism in an applied context. A blanket justification that it is aimed at understanding how all the bits and pieces in a system function is not good enough: again a very large programme of work will be necessary to cover them all, and there is no guarantee that one of the most vital cogs in the whole machine will not be among the last to be investigated.

Within a particular specialist discipline it is not difficult to see how mathematical models can assist in formulating hypotheses in a precise and convenient way; further, they can provide a useful framework for investigating the components in a systematic fashion. But agriculture, like ecology, is essentially multidisciplinary. Although it may be more difficult, the application of mathematical models in multidisciplinary studies, their strategic use in Mr Jeffers' words, has a much greater potential reward than within single disciplines. In agricultural research, models could form the essential link between the agronomist in the field and the specialist in the laboratory. The field man must split up his black box and find out what is going on within it: the specialist needs a more realistically defined purpose than simply understanding any or all of the mechanisms involved in isolation. Both these needs could be met by a common mathematical model, providing a common framework for the development of a viable team effort.

I believe we are just beginning to appreciate the strategic use of mathematical models in agricultural research to make multidisciplinary projects possible. As an agricultural ecologist, my definition of the challenge of mathematics is 'Can we make use of modelling in a multidisciplinary context?' Shall we be bold enough to change our ways of thought, and our organizational structures as necessary, to take advantage of this opportunity?

I am encouraged by the enthusiasm of the participants, and the valuable exchanges of ideas at this symposium with its ecological theme, which I translate as a multidisciplinary approach.

Self-assessment answers and comments

Question 1

They are not highly organized; lack of organization implies *high entropy*.

Question 2

$$\text{Sugar} + \text{oxygen} \longrightarrow \text{water} + \text{carbon} + \text{energy,}$$
$$\text{dioxide}$$

$$C_6H_{12}O_6 + 6O_2 \longrightarrow 6H_2O + 6CO_2 + X \text{ joules.}$$

Question 3

Sun; autotrophs; heterotrophs; environment; space.

The heterotroph stage may be missed out, with energy being lost directly from an autotroph to its environment.

Question 4

Components: autotrophs, heterotrophs, non-living components.

Energy transfer processes and material transfer processes (these are effectively the same): photosynthesis, tissue respiration, synthesis; nutrient intake, waste output, growth, reproduction, death.

Question 5

Conventional science	Dale's terminology
Observation	not included
Classification	Lexical phase; part of parsing
Experimentation	Analysis (?)
Synthesis (?)	Modelling

Clearly, there are similarities between the two approaches, but the order appears to differ.

Question 6

It does not attempt to classify the system in the terms used by Ackoff, or to define the purpose/goal/ideal of the system. This difference will be discussed in section 2.3.2.

Question 7

Autotrophs; heterotrophs; non-living materials. If you have any doubts, look back to section 1.3.

Question 8

Energy, as in measurements of flow. Numbers of organisms, as in population dynamics.

Question 9

If a quantitative model of the system can be constructed, it could be used to *simulate* the effect of the required input without the need to interfere with the entire ecosystem, provided that the model was known to be accurate.

Question 10

Population dynamics (see section 2.2).

Question 11

It is increasing at $3 - 2 = 1 \, \text{dm}^3 \, \text{min}^{-1}$.

Question 12

$$\Delta S = -\frac{S-P}{10}; \qquad \Delta P = +\frac{S-P}{10}.$$

The minus sign in the equation for ΔS indicates that this is a loss from the soil and the plus sign for ΔP indicates that it is an input into the plant.

$$S = 15.7; \qquad P = 14.3.$$

Question 13

$$\Delta X = kX_1, \qquad \text{where } X_1 \text{ is the value of } X \text{ at } T_1.$$

Question 14

$$f' = kw$$

Question 15

The function f' contains w, which is the dependent variable whose value is unknown.

Question 16

Surely you do not need an answer!

Question 17

Clearly, an animal cannot age by more than one day in one day and it cannot age backwards, so a transition from age 3–4 to 2–3, for example, has zero probability.

Question 18

	Home	Work
Home	$\frac{2}{7}$	$\frac{5}{7}$
Work	$\frac{5}{7}$	0

In this, the home–home and work–work transitions represent the probabilities of his staying at home for 24 hours and of his staying at work for 24 hours.

Question 19

(a) Leslie matrix.

(b) Differential equation.

(c) Compartment model, in conjunction with differential or difference equations.

Question 20

Because this agreement could have arisen purely *by chance* and there might to no real connection between the model and reality.

Question 21

$$\frac{23}{100} = 0.23.$$

Question 22

One way would be to observe the behaviour of the system on a number of occasions under similar conditions and to record how often each particular set of values occurred.

Question 23

The constant k.

Question 24

Because we should be effectively testing the experiments against themselves.

Question 25

Uniqueness of some systems; large number of variables; danger of damage to system; cost of experimentation etc.

Question 26

Why are we studying the system? See section 2.3.1 and earlier modules.

Question 27

Yes, it receives an input of solar energy.

Question 28

Grass; sheep; earthworms; slugs; snails; soil organic matter; soil minerals; air; water; etc.

Question 29

(a) Yes; solar energy taken up by the grass, passed on to the sheep, lost as heat from the sheep, removed in the wool, etc.

(b) Yes; by many of the processes listed in (a).

Question 30

Yes; we ought to inquire whether, as with other systems controlled by man, our system has a purpose or goal.

Question 31

Grass; sheep; man; soil (see Figure 17).

Question 32

Autotrophs: grass; heterotrophs: sheep, man; non-living: most of the soil, food store.

Question 33

250 g of grass per metre squared per year; 50 g of grass per metre squared per year, approximately.

Question 34

Since growth occurs continuously, a model using differential equations would be most appropriate.

Question 35

$$\frac{dw}{dt} \propto w \times l \qquad \text{or} \qquad \frac{dw}{dt} = k \times w \times l,$$

where k is some constant.

Question 36

$$\frac{dw}{dt} = 0.$$

Question 37

Symbiosis (see list in section 1.4.1).

Question 38

A transition matrix or Leslie matrix (see section 3.5).

Question 39

0.1.

Question 40

An economic system; man as a socio-technical system.

Question 41

Solar energy; cash; fertilizers; water. These are all physical inputs, but in addition we should also remember that there will be social inputs, such as education, the aspirations of the men involved in the system, and other social amenities. These all come from outside the system, but affect its operation, so they should also be included in the system's inputs.

Question 42

Cost; possible lack of replications; risk of damage to unique systems.

Index

order of differential equations	3.4
parameter	2.3.1; 2.3.6
peat	5.6.6
photosynthesis	1.3
population dynamics	2.2
probabilistic	4
probability	3.5; 4
rare ecosystems	2.3.7
rate	3.2
regeneration	5.6.1
regulation	3.2.1
replication	4
response curve	5.6.1
rumen	5.6.4
ruminant	5.6.4
selective breeding	7
selective pressures	7
separation of variables	3.4
simulation	4
solving differential equations	3.4
stable age structure	3.5
static properties of ecosystems	2.2
statistical tests	4
stem	5.6.1
stores	5.6.4
sward	5.6.1
systems environment	5.3
systems goals	5.4
tissue respiration	1.3
transition	3.5
upland areas	7
validation	4
variable	2.3.5

Acknowledgements

Grateful acknowledgement is made to the following for material used in this module:

Appendix: Blackwell Scientific Publications Ltd for N. R. Brockington 'Summary and assessment: an agricultural scientist's view' in J. N. R. Jeffers (ed.) *Mathematical Models in Ecology. 12th Symposium of the British Ecological Society* 1972.

Figure 3: Camera Press Ltd; *Figures 4 and 31*: *Farmers' Weekly*; *Figures 5 and 7*: Heather Angel; *Figure 28*: The Nature Conservancy; *Figure 29*: John Topham, Sidcup; *Figures 30 and 32*: Commercial Camera Craft, Yeovil; *Figure 34*: Crown Copyright, Central Veterinary Laboratory; *Figure 35*: Oxo Meat Cookery Service; *Figure 36*: International Wool Secretariat; *Figures 37 and 38*: drawn from unpublished data supplied by R. V. Large, Grassland Research Institute.

NOTES

NOTES

NOTES

Systems Behaviour

Module 1   Deep-sea container ports – Systems appraisal and simulation modelling

Module 2   Air traffic control – A man–machine system

Module 3   Industrial social systems

Module 4   A local government system – A case study of Brighton Corporation and the Brighton Marina

Module 5   The British telephone system

Module 6   The structure and management of ecosystems

Module 7   The human respiratory system

Module 8   A shipbuilding firm